Radio Systems Engineering

Héctor J. De Los Santos • Christian Sturm
Juan Pontes

Radio Systems Engineering

A Tutorial Approach

 Springer

Héctor J. De Los Santos
Irvine
California
USA

Christian Sturm
Bietigheim-Bissingen
Germany

Juan Pontes
Stuttgart
Germany

ISBN 978-3-319-34740-0   ISBN 978-3-319-07326-2 (eBook)
DOI 10.1007/978-3-319-07326-2
Springer Cham Heidelberg New York Dordrecht London

*Este libro lo dedico a mis queridos padres
y a mis queridos Violeta, Mara, Hectorcito,
y Joseph.*

Héctor J. De Los Santos

Para Yoyis y Suki

Juan Pontes

*"Y sabemos que a los que aman a Dios todas las cosas les ayudan a
bien, esto es, a los que conforme a su propósito son llamados."*

Romanos 8:28

Preface

This book is the outgrowth of the course "Modern Radio Systems Engineering," developed and taught by the Dr. H.J. De Los Santos during the 2010/2011 Winter Semester, while he held a German Research Foundation *Mercator Visiting Professorship* at the Institute for High-Frequency Engineering and Electronics (IHE), Karlsruhe Institute of Technology/University of Karlsruhe (TH), Germany. In the German system, the courses consist of lectures and exercises; there is no homework! Both lectures and exercises are held once a week for one and one-half hours. The teaching process entails the literal transfer of information to the student, meaning that all the material required to be learned in a given course must be explicitly given to the student in lecture notes handouts. The American concept of urging/demanding that students learn material by themselves, e.g., "read this chapter on your own," "convince yourself by deriving a result by yourself," etc., is nonexistent. This book attempts to recreate this learning experience. In particular, the same team who carried out the lecturing and tutorial exercises in the above course, namely, Dr. H.J. De Los Santos, and former Research Associates (now Drs.) Dr.-Ing. C. Sturm and Dr. Ing. J. Pontes, respectively, has reunited once more to collaborate in producing this book.

The course is aimed at upper-level undergraduates/first-year graduate students, who already have knowledge of devices and circuits for radio-frequency (RF) and microwave communications and are ready to study the systems engineering-level aspects of modern radio communications systems. In particular, the course gives a general overview of radio systems, together with their components. In this context, the focus is on the analog parts of the system, with their non-idealities. Based on the physical functionality of the various building blocks of a modern radio system, block parameters are derived, which allows the examination of their influence on the overall system performance.

The chapters of the book are complemented by tutorial exercises, based on the Keysight SystemVue electronic system-level (ESL) design software. In these tutorials, the readers gain practical experience with slightly simplified real-world design examples of radio systems, both in the area of communications as well as radar sensing. The tutorials cover state-of-the-art system standards and applications and consider the characteristics of typical radio-frequency hardware components. For all tutorials, a comprehensive description of the tasks, including some hints to the solutions, is provided. The readers are then intended to perform these tasks independently. Then, complete simulation models and solutions to the tutorial exercises is given.

Radio Systems Engineering: A Tutorial Approach fills a niche not addressed by previous books; in fact, it would make an excellent prerequisite for them. For instance, previous books are typically aimed at advanced graduate students and practicing engineers. As a result, they tend to be specialized by providing an in-depth focus on specific applications and technologies, so as to be a resource to individuals developing these applications. This book, on the other hand, facilitates the integration of the knowledge gathered by undergraduate students during their pre-Senior years, which primarily focuses on devices and circuits, and gives them their first exposure to their exploitation in engineering an overall system to fulfill real-life performance requirements.

Radio Systems Engineering: A Tutorial Approach contains nine chapters. Chapter 1 starts by providing an overview of wireless communication systems, defining the fundamental wireless communications problem and motivating approaches to its solution. Chapter 2 introduces system-level block diagrams of Amplitude Modulation (AM) and Frequency Modulation(FM)/Phase Modulation (PM) modulators and demodulators, and explains their respective principles of operation. Chapter 3 addresses a number of topics surrounding system performance parameters as well as a system-level description of component models for systems analysis. Chapter 4 discusses the fundamentals of the radio channel and a discussion of antenna parameters. Chapter 5 deals with the topics of Noise, Nonlinearity and Time Variance as it pertains to relating the performance of building blocks to that of the overall system built as a cascade of them. Chapter 6 focuses on the topics of sensitivity and dynamic range for a receiver, including how the performance of the individual building blocks impacts the overall system. Chapter 7, addresses the topics of transmitter and receiver architectures, and practical aspects impacting the performance of oscillators. Chap. 8 integrates the knowledge presented in Chaps. 1–7 by engaging into a "case study" tutorial discussion that exposes the engineering considerations and thought processes behind the design of two real-life receivers, implemented after the WCDMA and LTE standards, respectively. Chapter 9, includes five tutorial exercises based on the Keysight SystemVue software tool, a free limited- time license of which may be downloaded from http://www.keysight.com/find/eesof-radio-systems-engineering.

Acknowledgments

Prof. Dr.-Ing. Thomas Zwick, Director of the Institute for High-Frequency Engineering and Electronics (IHE), Karlsruhe Institute of Technology/University of Karlsruhe (TH), Germany, and his immediate predecessor, Prof. Dr.-Ing. Dr.h.c. Dr.-Ing. E.h. Werner Wiesbeck, are greatfully thanked by Dr. H.J. De Los Santos, for being instrumental in enabling his teaching and research stay at IHE, KIT as German Research Foundation (DFG) Mercator Visiting Professor during 2010–2011, the context within which most of the material for the book was put together. He also greatfully thanks Prof. Dr.-Ing. Zwick for his essential role in conceiving the outline of the course "Modern Radio Systems Engineering," which was the basis for preparing the lecture notes which have been developed into this book. Dr. De Los Santos gratefully acknowledges the DFG for its support under the *Mercator Visiting Professorship* Award. The key assistance of Mr. Greg Oldham, Field Engineer, Keysight Technologies, in facilitating the SystemVue and ADS software tools used in developing the tutorials is especially acknowledged. In relation to these tools, the enthusiastic assistance of Mr. John Kikuchi, Keysight Technologies EEsof University Program Manager, Mr. Frank Ditore, Product Planning & Marketing Manager, Mr. Brian Buchanan, Contracts, and Mr. Dan Winter, EEsof EDA, all of Keysight Technologies, is greatfully acknowledged. The assistance of the staff at Springer US is gratefully acknowledged, in particular, Mr. Charles B. Glaser, Executive Editor, for facilitating the opportunity to work in this project, and Ms. Jessica Lauffer, Assistant Editor, Engineering, for her assistance pertaining to manuscript development.

April, 2014

Héctor J. De Los Santos
Christian Sturm
Juan Pontes

Contents

Chapter 1
Introduction to Radio Systems

Abstract In this chapter, an overview of wireless communication systems is presented. We begin with a definition of the fundamental wireless communications problem and motivate approaches to its solution. Following definitions of carrier, baseband, and modulation, the simplified block diagram of a transmitter, together with a description of its typical building block elements, is presented. We then focus on the mathematical description and corresponding spectral properties of amplitude modulation (AM), frequency modulation (FM), and phase modulation (PM). This is followed by a study of differences among modulation schemes, their bandwidth, and their vectorial representation. In particular, the separate cases of narrow- and wideband FM, and phase modulation (PM) are addressed.

1.1 Overview of Wireless Communication Systems

The fundamental problem of wireless communications consists in transferring information between a *source* and a *destination*, Fig. 1.1 [1, 2].

If the signal representing the information to be transmitted is in *analog* form, i.e., exists at all times, a straightforward way to accomplish this information transfer would be to feed this signal to an antenna (at the source), which will convert it into an electromagnetic (EM) wave and radiate it into space. If the signal at the source represents, for example, *music or voice*, which have a maximum frequency content of about $f = 20\ kHz$, this would mean that, for reasonable efficiency, the transmitting antenna required would have dimensions of half-wavelength given by,

$$\frac{\lambda}{2} = \frac{c}{2f} = \frac{3\times10^{8}\,m/s}{2(20\times10^{3}/s)} = 7000m = 4.35mi \qquad (1.1)$$

where λ is the wavelength and c is the speed of light. Obviously, the required antenna length of this example, namely, 4.35 miles, would be impractical. This state of affairs is circumvented in practice by the process of *modulation*.

Modulation is the *process* by which the information whose transmission is desired is impressed upon a *high-frequency signal*. The set of frequencies that comprise the

H. J. De Los Santos et al., *Radio Systems Engineering,*
DOI 10.1007/978-3-319-07326-2_1, © The Authors 2015

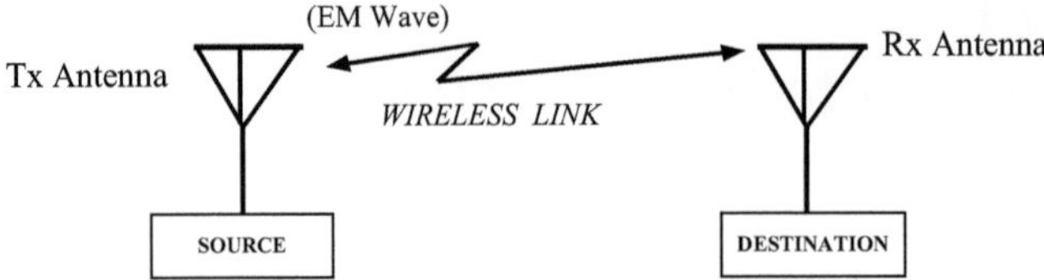

Fig. 1.1 Depiction of the wireless communications problem. *Tx* is transmitter; *EM* is electromagnetic; *Rx* is receiver; *Source* is the place of transmission; *Destination* is the place of reception

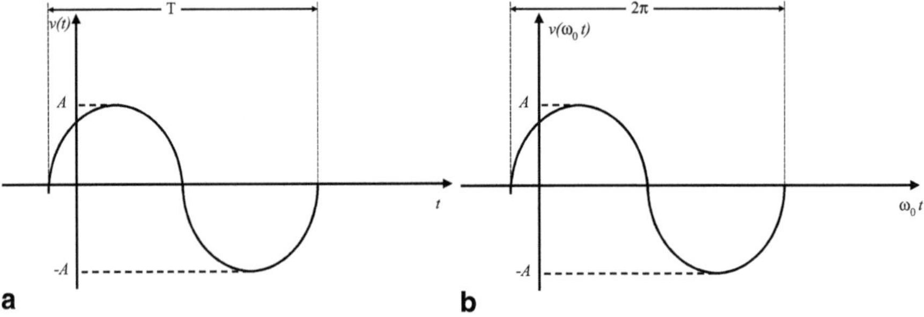

Fig. 1.2 Depiction of carrier signal. **a** As function of time. **b** As function of phase

desired information to be transmitted, $m(t)$, in its original form, is referred to as the *baseband* signal. On the other hand, the continuous high-frequency wave which carries the baseband signal is referred to as the *carrier*. The modulation of the carrier is effected when the amplitude, the frequency, or the phase of the carrier is made to vary in a manner dictated by the baseband signal.

The general equation for a continuous single-frequency time function is given by,

$$v(t) = A\cos(\omega_0 t + \varphi_0) = A\cos\left[\varphi(t)\right] \qquad (1.2)$$

which may be represented as the function of time or phase $v(t)$ as depicted in Fig. 1.2.

The parameters in (1.2) embody the following. The factor A quantifies the peak amplitude of $v(t)$, the parameter ω_0 is called the radial frequency of the time function described by (1.2), and the time, T, shown in Fig. 1.2a, is called the *period* of the cosine function; it is the time required to change the phase of the time function by 2π radians. Differentiation of (1.2) with respect to time yields the fundamental relationship,

$$\omega_0 = \frac{d\varphi}{dt} \qquad (1.3)$$

where, $\varphi = \omega_0 t + \varphi_0$. From Fig. 1.2 and (1.2) it is seen that a time change of T changes the phase by 2π, i.e.,

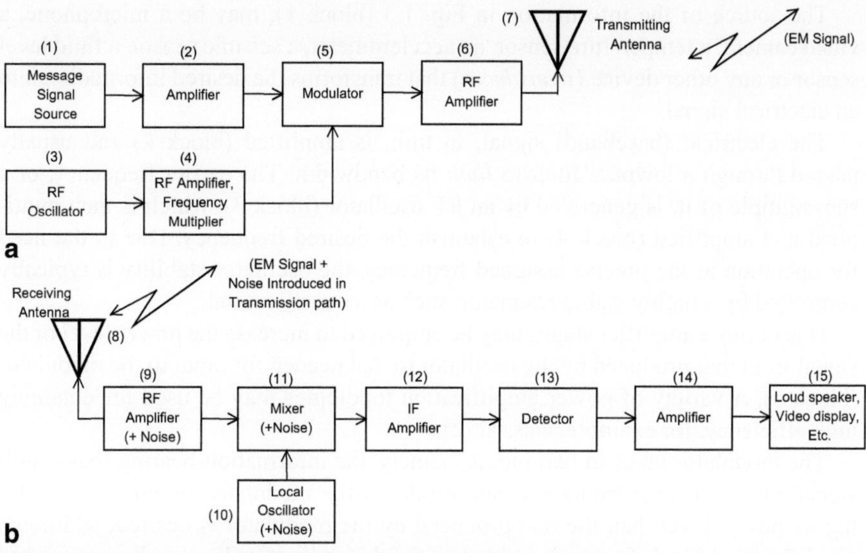

Fig. 1.3 Simplified block diagram of **a** transmitter, and **b** of a receiver

$$\omega_0(t+T)+\varphi_0-(\omega_0 t+\varphi_0)=2\pi \tag{1.4}$$

from where one obtains the relationship,

$$\omega_0 T = 2\pi \tag{1.5}$$

or

$$T = \frac{2\pi}{\omega_0} = \frac{1}{f_0} \tag{1.6}$$

where f_0 is the frequency of the continuous wave (CW) measured in *Hertz*. The relation between radial and cyclical frequency is given by,

$$f_0 = \frac{\omega_0}{2\pi} \tag{1.7}$$

1.2   Simplified Block Diagram of Transmitter and Receiver

In this section we begin elaborating on the constituents of the fundamental wireless communications system of Fig. 1.1. In particular, Fig. 1.3 depicts a simplified transmitter and receiver system [1, 2].

The source of the information in Fig. 1.3 (block 1), may be a microphone, a video camera, a temperature sensor, an accelerometer, a seismic sensor, a fluid level sensor or any other device (*transducer*) that transforms the desired information into an electrical signal.

The electrical (baseband) signal, in turn, is amplified (block 2) and usually passed through a lowpass filter to *limit* its bandwidth. The carrier frequency, or a sub-multiple of it, is generated by an RF oscillator (block 3), which is then multiplied and amplified (block 4) to establish the desired frequency. Due to the need for operation at the precise assigned frequency, the oscillator stability is typically controlled by a highly stable resonator, such as a quartz crystal.

One or more amplifier stages may be employed to increase the power level of the signal from that produced by the oscillator to that needed for input to the modulator (block 5). A variety of power amplification topologies may be used for obtaining high efficiency, for example, class C [2].

The modulator takes in two inputs, namely, the information-bearing (baseband) signal and the carrier frequency, and produces the modulated output carrier. If a higher power level than the one produced by the modulator is desired, additional amplification (block 6) may be added so the desired power level to be transmitted by the antenna (block 7) is reached.

On the other hand, the transmitted information is captured by a receiver, Fig. 1.3b. The receiving antenna (block 8) utilized may be *omnidirectional* for general service, or *highly directional* for point-to-point communication. The received *wave induces* a small voltage in the receiving antenna, with amplitudes ranging from tens of millivolt to less than a microvolt, depending on a wide variety of conditions, in particular, the nature of the intervening transmitter-receiver space or channel. Notice that, en-route to the receiver, the transmitted signal picks up *noise* from the environment; this refers to random signals which alter the amplitude, phase, or frequency of the transmitted carrier. The received signal delivered by the antenna is *amplified by a low noise amplifier* (LNA) (block 9) to increase the signal power to a level appropriate for input to a mixer. The LNA also provides isolation between the local oscillator (LO) (block 10) and the antenna, as well as increasing the received signal amplitude to overcome the noise that is inevitably introduced in the mixer (block 11). The mixer is a nonlinear circuit which produces multiples of the sum and difference of the RF and local oscillator (LO) signal frequencies, thus frequency-translating the received carrier signal, f_{RF}, to the intermediate frequency, f_{IF}, where demodulation is to be effected. In the receiver architecture shown, the LO is tuned to produce a frequency that differs from the incoming signal frequency f_{RF} by the intermediate frequency f_{IF}; in other words, f_{LO} can be equal to either $f_{RF}+f_{IF}$ or $f_{RF}-f_{IF}$. The IF amplifier (block 12) increases the signal amplitude to a level appropriate for detection, and provides most of the frequency selectivity necessary to "pass" the desired signal and reject the undesired signals that are found in the mixer's output spectrum. The detector (block 13) extracts the original message from the modulated IF input. The extracted signal is amplified (block 14) to an amplitude that is appropriate for driving a loudspeaker, a television tube, or other output device. The output transducer (15) converts the signal information back to its original form, e.g., a sound wave, a picture, etc.

1.3   Basic Modulation Definitions in Mathematical Terms

In the process of modulation, a property of the carrier signal is varied under the influence of the baseband signal. In particular, given the mathematical representation of the carrier by,

$$v(t) = A\cos(\omega_0 t + \varphi_0) = A\cos[\varphi(t)] \tag{1.8}$$

three parameters in this equation may be made independently time-varying, namely, the amplitude A, giving rise to Amplitude Modulation (AM), the frequency, giving rise to Frequency Modulation (FM), or the phase, giving rise to Phase Modulation (PM). To make these possible variations explicit in (1.8), we begin by assuming that the radial frequency, rather than being a constant, is a function of time, i.e.,

$$\omega = \frac{d\varphi}{dt} \tag{1.9}$$

which, when integrated, gives as the phase of the cosine wave,

$$\varphi = \int_0^t \omega dt + \varphi_0 \tag{1.10}$$

Permitting also the amplitude A to be a time-dependent variable, one obtains, following substitution of (1.10) into (1.8),

$$v(t) = A(t)cos\left(\int \omega dt + \varphi_0\right) \tag{1.11}$$

We now address the mathematical representation of each of these modulation schemes, AM, FM and PM [1].

1.3.1   Amplitude Modulation (AM)

If, in (1.11), the radial frequency is held constant and the amplitude, $A(t)$, allowed to vary in the manner given by,

$$A(t) = a_0\left[1 + mg(t)\right] \tag{1.12}$$

then, the time function $v(t)$ is said to be *amplitude modulated.* In (1.12), a_0 is a constant, m is the *modulation index*, and $g(t)$ is the *modulation function.*

1.3.2   Frequency Modulation (FM)

In frequency modulation, as its name implies, one holds the amplitude A constant and allows the radial frequency to vary in a manner given by,

$$\omega(t) = \omega_0 \left[1 + mg(t)\right] \tag{1.13}$$

Here, ω_0 is a constant, m is the modulation index, and $g(t)$ is the modulation function.

1.3.3   Phase Modulation (PM)

In phase modulation, the *envelope A* is held constant and the phase is allowed to vary in the fashion dictated by,

$$\varphi(t) = \omega_0 \left[1 + \varphi_0 mg(t)\right] \tag{1.14}$$

1.4   Spectral Properties of the Basic Modulation Schemes

In addition to enabling the transmission of information by a relatively small antenna, the process of modulating a high-frequency carrier signal yields peculiar spectral characteristics; these are presented next.

1.4.1   Amplitude Modulation Spectrum

For simplicity, we first consider a baseband modulation function, $g(t)$, that consists of a pure cosinusoid, i.e.,

$$g(t) = \cos p_1 t \tag{1.15}$$

where p_1 is the baseband radial frequency. Successive substitutions of (1.15) into (1.12) into (1.11) yields (with $\omega = \omega_0$, $\phi_0 = 0$),

$$v(t) = a_0 \left[1 + m\cos p_1 t\right]\cos \omega_0 t \tag{1.16}$$

If one now uses the trigonometric identity,

$$\cos x \cos y = \frac{1}{2}\left[\cos(x+y) + \cos(x-y)\right] \tag{1.17}$$

one has, in place of (1.17),

$$v(t) = a_0 \left[\cos \omega_0 t + \frac{m}{2}\cos(\omega_0+p_1)t + \frac{m}{2}\cos(\omega_0-p_1)t\right] \tag{1.18}$$

From (1.15) and (1.18), we find that the process of amplitude modulation has *changed* the radian frequency from baseband, p_1, to two *sidebands*, ω_0+p_1, ω_0-p_1,

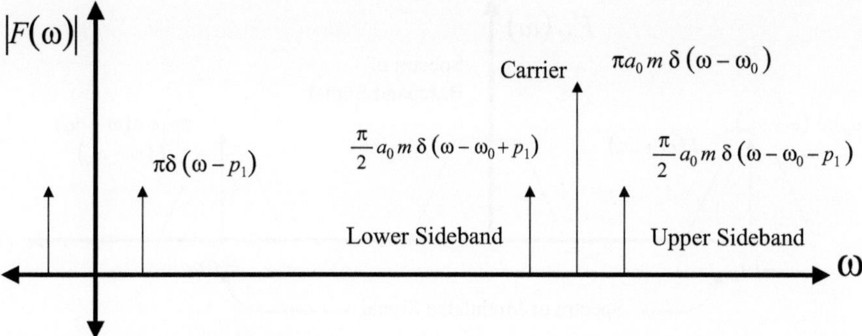

Fig. 1.4 Spectrum of amplitude modulated single-frequency signal. Whereas the single-frequency carrier has a spectrum represented by the delta function $\pi\delta(\omega - p_1)$, the spectrum of the modulated carrier has a spectrum represented by $\pi a_0 m\delta(\omega - \omega_0)$, the lower sideband $\dfrac{\pi}{2}a_0 m\delta(\omega - \omega_0 + p_1)$, and the upper sideband $\dfrac{\pi}{2}a_0 m\delta(\omega - \omega_0 - p_1)$

centered at the carrier radial frequency, ω_0. This results in the spectrum depicted in Fig. 1.4.

Rather than a single cosinusoid, let us next assume that the baseband modulation function is *general* in form. In this case, taking its Fourier transform results in,

$$F_B(\omega) = \int_{-\infty}^{\infty} mg(t)e^{-j\omega t}\,dt \tag{1.19}$$

On the other hand, for the Fourier transform of the modulated time function we obtain, similarly,

$$F_M(\omega) = a_0 \int_{-\infty}^{\infty} \left[1 + mg(t)\right]\cos \omega_0 t\, e^{-j\omega t}\,dt \tag{1.20}$$

Upon expansion, this yields,

$$F_M(\omega) = \frac{a_0}{2}\left[2\pi\delta(\omega - \omega_0) + 2\pi\delta(\omega + \omega_0) + m\int_{-\infty}^{\infty} g(t)\left\{e^{-j(\omega-\omega_0)t} + e^{-j(\omega+\omega_0)}\right\}\right]dt. \tag{1.21}$$

$$F_M(\omega) = \frac{a_0}{2}\left[2\pi\delta(\omega - \omega_0) + 2\pi\delta(\omega + \omega_0) + F(\omega - \omega_0) + F(\omega + \omega_0)\right] \tag{1.22}$$

where,

$$F(\omega \mp \omega_0) = m\int_{-\infty}^{\infty} g(t)e^{-j(\omega \mp \omega_0)t}\,dt \tag{1.23}$$

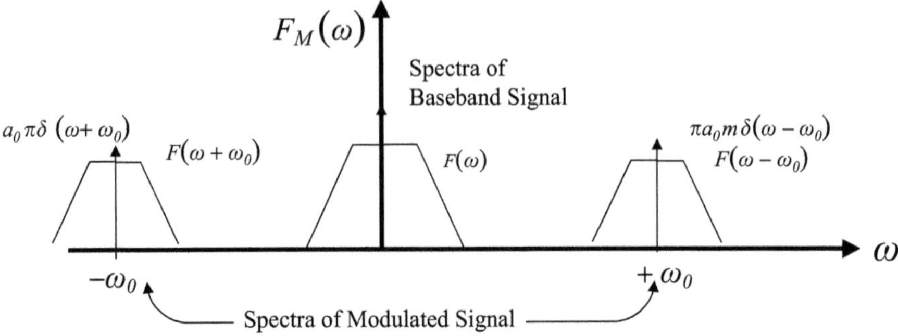

Fig. 1.5 Spectrum of amplitude modulated baseband signal

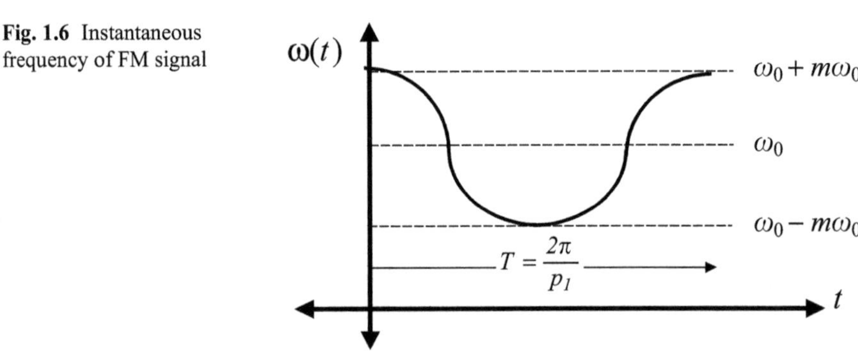

Fig. 1.6 Instantaneous frequency of FM signal

represents a shift in frequency from baseband to a frequency near the carrier. This results in the spectrum depicted in Fig. 1.5.

1.4.2　Frequency Modulation Spectrum

We begin our discussion of the mathematical properties of FM by writing the FM time function in its general form,

$$v(t) = A\cos\left[\int \omega_0(1+mg(t))dt + \varphi_0\right] \tag{1.24}$$

and then, restricting the modulation function to a single cosinusoid, with $\varphi_0 = 0$, (1.24) becomes,

$$v(t) = A\cos\left[\omega_0 t + \frac{m\omega_0}{p_1}\sin p_1 t\right] \tag{1.25}$$

For this special case, a plot of *instantaneous* frequency, $\omega = \dfrac{d\varphi}{dt}$, yields Fig. 1.6.

The ratio of the maximum deviation in instantaneous frequency, $\Delta\omega_{MAX}$, to the input modulating frequency, p_1, is called the *modulation index*, β or M_p. From Fig. 1.6 and (1.25), we have,

$$\beta \equiv \frac{\Delta\omega_{MAX}}{p_1} = \frac{m\omega_0}{p_1} = M_p \tag{1.26}$$

The substitution of (1.26) into (1.25) gives,

$$v(t) = A\cos\left[\omega_0 t + M_p \sin p_1 t\right] \tag{1.27}$$

If we now expand (1.27), making use of the trigonometric identity,

$$\cos(x+y) = \cos x \cos y - \sin x \sin y \tag{1.28}$$

we obtain, in place of (1.27),

$$v(t) = A\cos \omega_0 t \cos(M_p \sin p_1 t) - A\sin \omega_0 t \sin(M_p \sin p_1 t) \tag{1.29}$$

A generic function for ordinary *Bessel functions* of the first kind is given by,

$$e^{\pm jM_p \sin p_1 t} = J_0(M_p) + 2\sum_{k=1}^{\infty} J_{2k}(M_p)\cos 2kp_1 t \pm 2j\sum_{k=0}^{\infty} J_{2k+1}(M_p)\sin(2k+1)p_1 t$$

$$\tag{1.30}$$

This, in conjunction with the exponential equations for the sine and cosine functions, $\sin x = \dfrac{e^{jx} - e^{-jx}}{2j}$ and $\cos x = \dfrac{e^{jx} + e^{-jx}}{2}$, allows us to write, with, $x = M_p \sin p_1 t$,

$$\sin(M_p \sin p_1 t) = 2\sum_{k=0}^{\infty} J_{2k+1}(M_p)\sin(2k+1)p_1 t \tag{1.31}$$

and

$$\cos(M_p \sin p_1 t) = J_0(M_p) + 2\sum_{k=1}^{\infty} J_{2k}(M_p)\cos(2k)p_1 t \tag{1.32}$$

Substitution of (1.31) and (1.32) into (1.29) yields,

$$v(t) = A\cos \omega_0 t \left[J_0(M_p) + 2\sum_{k=1}^{\infty} J_{2k}(M_p)\cos 2kp_1 t \right] -$$
$$A\sin \omega_0 t \left[2\sum_{k=0}^{\infty} J_{2k+1}(M_p)\sin(2k+1)p_1 t \right] \tag{1.33}$$

If (1.33) is divided by the constant A, we obtain as the first few terms of the series,

$$\frac{v(t)}{A} = J_0(M_p)\cos \omega_0 t - 2J_1(M_p)\sin p_1 t \sin \omega_0 t$$
$$+ 2J_2(M_p)\cos \omega_0 t \cos 2p_1 t - 2J_3(M_p)\sin 3p_1 t \sin \omega_0 t$$
$$+ 2J_4(M_p)\cos \omega_0 t \cos 4p_1 t - 2J_5(M_p)\sin 5p_1 t \sin \omega_0 t \qquad (1.34)$$

If the product terms are now changed to sum and difference terms, we have,

$$\frac{v(t)}{A} = J_0(M_p)\cos \omega_0 t - 2J_1(M_p)\left[\cos(\omega_0 + p_1)t - \cos(\omega_0 - p_1)t\right]$$
$$+ J_2(M_p)\left[\cos(\omega_0 + 2p_1)t + \cos(\omega_0 - 2p_1)t\right] +$$
$$+ J_3(M_p)\left[\cos(\omega_0 + 3p_1)t - \cos(\omega_0 - 3p_1)t\right] +$$
$$+ J_4(M_p)\left[\cos(\omega_0 + 4p_1)t + \cos(\omega_0 - 4p_1)t\right] + . \qquad (1.35)$$

The comparison of (1.35) with (1.18) reveals certain differences, which we now explore.

1.4.3   Differences Between AM and FM

It has been shown, see (1.18), that amplitude modulation (AM) produces a *single* set of sidebands. An inspection of (1.35), however, reveals that FM, on the other hand, produces an *infinite* number of sidebands. Each sideband is *separated* from the carrier by a *frequency* kp_1, where k is an integer and p_1 is the modulating frequency. Another more subtle difference between these two forms of modulation relates to the partitioning of power between the carrier and the sidebands; this subtlety is now shown. The total average power contained in any time function is given by,

$$\bar{P}_T = \frac{1}{T}\int_{T\to\infty} |v(t)|^2 \, dt \qquad (1.36)$$

For the cosinusoidal AM wave given by,

$$v(t) = a_0\left[\cos \omega_0 t + \frac{a_0 m}{2}\cos(\omega_0 + p_1)t + \cos(\omega_0 - p_1)t\right] \qquad (1.37)$$

since the cosine function is orthogonal, the integral of cross-product terms obtained when (1.37) is squared, are zero. Therefore,

$$\bar{P}_T = \frac{a_0^2}{2} + \frac{a_0^2 m^2}{8} + \frac{a_0^2 m^2}{8} \qquad (1.38)$$

Table 1.1 Amplitudes of FM wave components

$\Delta\omega_M$	p_1	M_p	$J_0(M_p)$	$J_1(M_p)$	$J_2(M_p)$	$J_3(M_p)$	$J_4(M_p)$	$J_5(M_p)$	$J_6(M_p)$	$J_7(M_p)$
1000	3000	1/3	0.9725	0.1644	0.03					
1000	1500	1/2	0.8930	0.3138	0.06					
1000	1000	1	0.7652	0.4401	0.1149	0.0196				
1000	500	2	0.2339	0.5767	0.3528	0.1289	0.034			
1000	333	3	-0.2501	0.3391	0.4861	0.3091	0.1320	0.043		
1000	250	4	-0.3971	-0.06604	0.3641	0.4302	0.4302	0.1321	0.0491	
1000	100	5	-0.1776	-0.3276	0.1697	0.2404	0.3912	0.2611	0.1311	0.0533

where the first term represents the power in the carrier, and the second and third terms that in the lower and upper sidebands, respectively. From (1.38) we find that the power contained in the carrier is *independent* of the properties of the modulating function (captured by *m*). The situation is somewhat altered in the case of FM waveforms. For FM waves, since the envelope is constant, the total power becomes,

$$\bar{P}_T = \frac{A^2}{2} \tag{1.39}$$

It follows, from (1.39), that in an FM wave the total average power, carrier plus sidebands, is *constant*. If (1.35) is squared, and the orthogonality property used, we obtain the relationship,

$$J_0^2(M_p) + 2\sum_{k=1}^{\infty} J_k^2(M_p) = 1 \tag{1.40}$$

In (1.40), $J_0^2(M_p)$ is the *fraction* of total power in an FM wave that is contained in the carrier. Since $M_p = \dfrac{\Delta\omega_M}{p_1}$, the carrier power is clearly *dependent* on the properties of the modulating function. The following example illustrates some of the properties of a cosinusoidal FM wave. Let's take $\Delta\omega_M = 1000s^{-1}$, and allow p_1 to vary between 3000 and 200 s^{-1}. This produces the values shown in Table 1.1, and the histogram shown as Fig. 1.7.

It is observed that all significant sidebands are contained within a bandwidth, called the *FM bandwidth*, given by [1],

$$\omega_T \approx 2(\Delta\omega_M + p_1) \tag{1.41}$$

1.4.4 Vector Representations

It has been found that insight into the behavior and properties of the AM and FM modulation schemes may be attained by representing the pertinent waveforms by rotating phasors; these are introduced next.

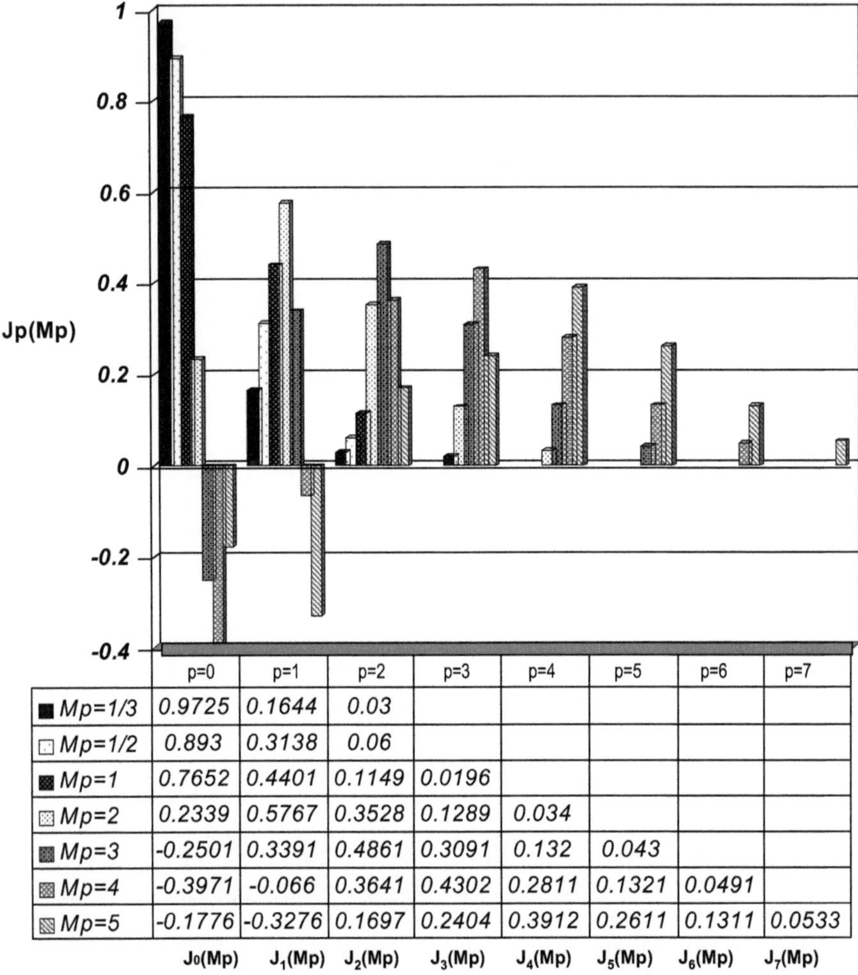

	p=0	p=1	p=2	p=3	p=4	p=5	p=6	p=7
■ Mp=1/3	0.9725	0.1644	0.03					
▫ Mp=1/2	0.893	0.3138	0.06					
▨ Mp=1	0.7652	0.4401	0.1149	0.0196				
▦ Mp=2	0.2339	0.5767	0.3528	0.1289	0.034			
▩ Mp=3	-0.2501	0.3391	0.4861	0.3091	0.132	0.043		
▨ Mp=4	-0.3971	-0.066	0.3641	0.4302	0.2811	0.1321	0.0491	
▨ Mp=5	-0.1776	-0.3276	0.1697	0.2404	0.3912	0.2611	0.1311	0.0533
	$J_0(Mp)$	$J_1(Mp)$	$J_2(Mp)$	$J_3(Mp)$	$J_4(Mp)$	$J_5(Mp)$	$J_6(Mp)$	$J_7(Mp)$

Fig. 1.7 Amplitude of FM sidebands (Bessel coefficients) as function of modulation index, M_p. p is the order of the Bessel function (0 through 7)

1.4.4.1 AM Vector Representation

We begin our treatment of this subject by considering the complex quantity, $e^{j\omega_0 t}$. In particular, we can think of this as a two-dimensional vector of unit amplitude rotating at an angular speed ω_0, where projection of the vector on the real axis is the cosine function, and its projection on the imaginary axis the sine function. This is shown in Fig. 1.8.

Let us now write the cosinusoidal AM wave as,

$$v(t) = Re\left\{a_0(1 + m\cos p_1 t)e^{j\omega_0 t}\right\} \tag{1.42}$$

Fig. 1.8 Vector representation of complex quantity $e^{j\omega_0 t}$

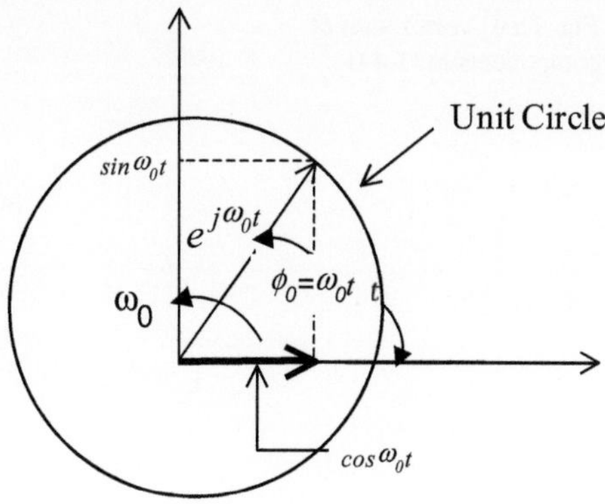

Fig. 1.9 Vector representation of (1.44) at time t_0

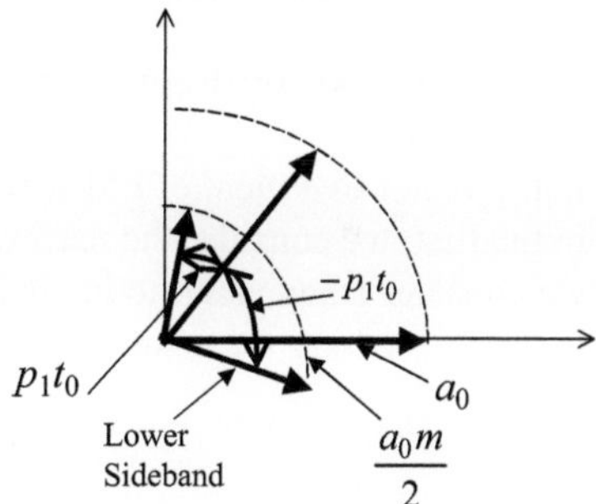

where *Re* signifies the "real part of." If the cosine term in (1.42) is written as,

$$\cos\, p_1 t = \frac{e^{jp_1 t} + e^{-jp_1 t}}{2} \tag{1.43}$$

and substituted back into (1.42), we have,

$$v(t) = Re\left\{ a_0 e^{j\omega_0 t} + \frac{a_0 m}{2} e^{j(\omega_0 + p_1)t} + \frac{a_0 m}{2} e^{j(\omega_0 - p_1)t} \right\} \tag{1.44}$$

From (1.44), we find that amplitude modulation with a cosinusoid produces two additional *vectors* with radial frequencies corresponding to upper and lower sidebands. Let us now freeze the motion of the vectors at some arbitrary time, t_0. Then, the lengths and directions would appear as shown in Fig. 1.9.

It is obvious, from (1.44) and Fig. 1.9, that the two vectors comprising the sidebands produce a composite vector that is in phase with the carrier. This is shown in Fig. 1.10.

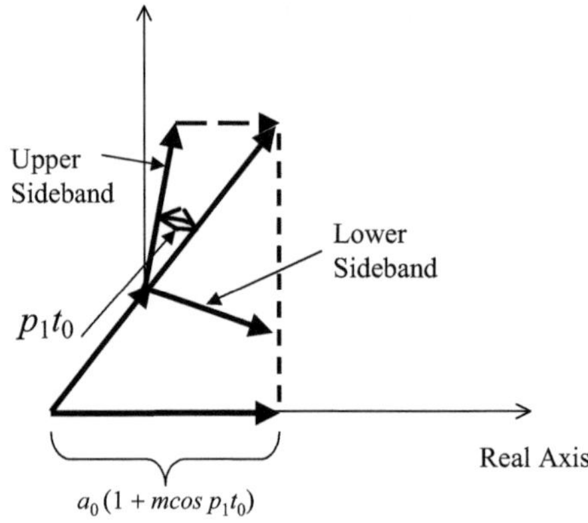

Fig. 1.10 Vector sum of components in (1.44)

1.4.4.2   FM Vector Representation

In the vector treatment of FM waveforms, we subdivide the analysis into two parts. In the first, we consider the narrowband FM case, in which $M_p \ll 1$. In the second, we consider frequency modulated waveforms with $M_p \sim > 1$. The treatment begins by, again, writing the cosinusoidally modulated FM waveform time function. This reads,

$$v(t) = A\cos\left[\omega_0 t + M_p \sin p_1 t\right]$$

$$= A\,Re\left\{e^{j(\omega_0 t + M_p \sin p_1 t)}\right\} \tag{1.45}$$

Since, $e^{jM_p \sin p_1 t} = \cos(M_p \sin p_1 t) + j\sin(M_p \sin p_1 t)$, we can write, in place of (1.45),

$$v(t) = ARe\left\{e^{j\omega_0 t}\left[\cos(M_p \sin p_1 t) + j\sin(M_p \sin p_1 t)\right]\right\} \tag{1.46}$$

1.4.4.2.1   Narrowband FM

With $M_p \ll 1$, we can expand the cosine and since functions of (1.46) by the first terms in a power series expansion, i.e.,

$$\cos(M_p \sin p_1 t) \cong 1$$
$$\sin(M_p \sin p_1 t) \cong M_p \sin p_1 t \tag{1.47}$$

Substitution of (1.47) into (1.46) yields,

$$v(t) \cong ARe\left\{e^{j\omega_0 t}\left[1 + jM_p \sin p_1 t\right]\right\} \tag{1.48}$$

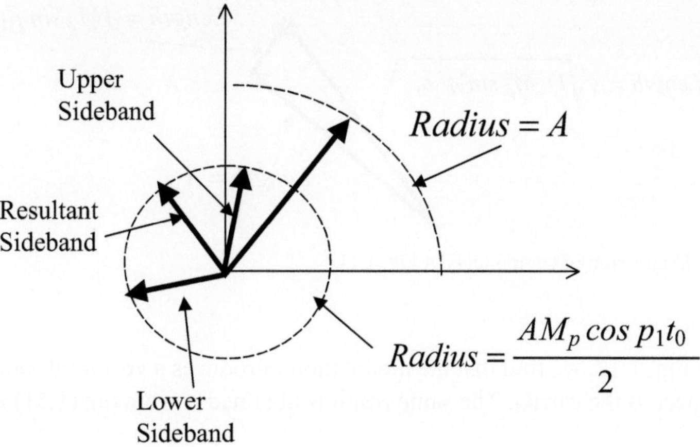

Fig. 1.11 Vectors of in (1.50)

The sine function of (1.48) can also be written as,

$$\sin p_1 t = \frac{e^{jp_1 t} - e^{-jp_1 t}}{2j} \tag{1.49}$$

which, upon substitution into (1.48) yields,

$$v(t) \cong A Re\left\{ e^{j\omega_0 t} + \frac{1}{2} M_p \left(e^{j(\omega_0 + p_1)t} - e^{j(\omega_0 - p_1)t} \right) \right\} \tag{1.50}$$

From (1.50), we find that with $M_p < < 1$, we only have one set of significant side-bands. However, unlike AM, the lower sideband is shifted by π radians. If we write (1.50) in the form,

$$v(t) = A \cos \omega_0 t + \frac{A}{2} M_p \left[\cos (\omega_0 + p_1)t - \cos(\omega_0 - p_1)t \right] \tag{1.51}$$

and compare (1.51) to (1.35), we obtain identical results if we assume,

$$J_k(M_p) \cong 0, \; k > 1.$$

$$J_1(M_p) = \frac{M_p}{2}. \tag{1.52}$$

$$J_0(M_p) = 1$$

Finally, if we freeze time at some arbitrary instant, t_0, the vectors of (1.50) are arranged as shown in Figs. 1.11 and 1.12.

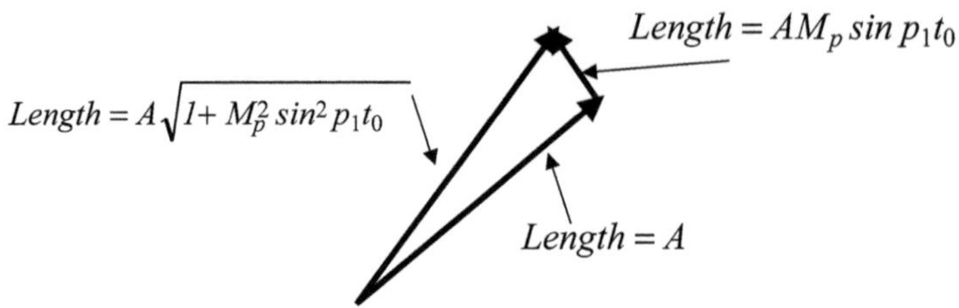

Fig. 1.12 Vector sum of components in Fig. 1.11

From Fig. 1.12, we find that the modulation introduces a vector 90° out-of-phase with respect to the carrier. The same result is obtained if we write (1.51) as,

$$v(t) = A\cos\omega_0 t - AM_p \sin \omega_0 t \sin p_1 t$$

$$= A\cos \omega_0 t + AM_p \sin p_1 t \cos(\omega_0 t + \pi) \tag{1.53}$$

The out-of-phase component produces a periodic speed-up and slow-down in the radial frequency, while the amplitude remains fairly constant, since,

$$A\sqrt{1 + M_p^2 \sin^2 p_1 t} \cong A \tag{1.54}$$

1.4.4.2.2 Wideband FM

Let us now remove the narrowband constraint. Then, from (1.33), repeated here for convenience,

$$v(t) = A\cos \omega_0 t \left[J_0(M_p) + 2\sum_{k=1}^{\infty} J_{2k}(M_p)\cos 2kp_1 t \right]$$

$$- A\sin \omega_0 t \left[2\sum_{k=0}^{\infty} J_{2k+1}(M_p)\sin(2k+1)p_1 t \right] \tag{1.33}$$

using,

$$Im\{Z\} = Re\{-jZ\} \tag{1.55}$$

we have,

$$v(t) = Re\left\{ Ae^{j\omega_0 t} \left[J_0(M_p) + 2\sum_{k=1}^{\infty} J_{2k}(M_p)\cos 2kp_1 t + 2\sum_{k=1}^{\infty} J_{2k+1}(M_p)\sin(2k+1)t \right] \right\} \tag{1.56}$$

As shown in (1.56), the even order sidebands (those with coefficient $J_{2k}(M_p)\cos 2kp_1 t$), are in phase with the carrier, while all the odd order harmonics lead the carrier by a constant phase of 90°. We next illustrate: (1) How the phase deviates from the carrier by an amount that is dependent on the phase of the

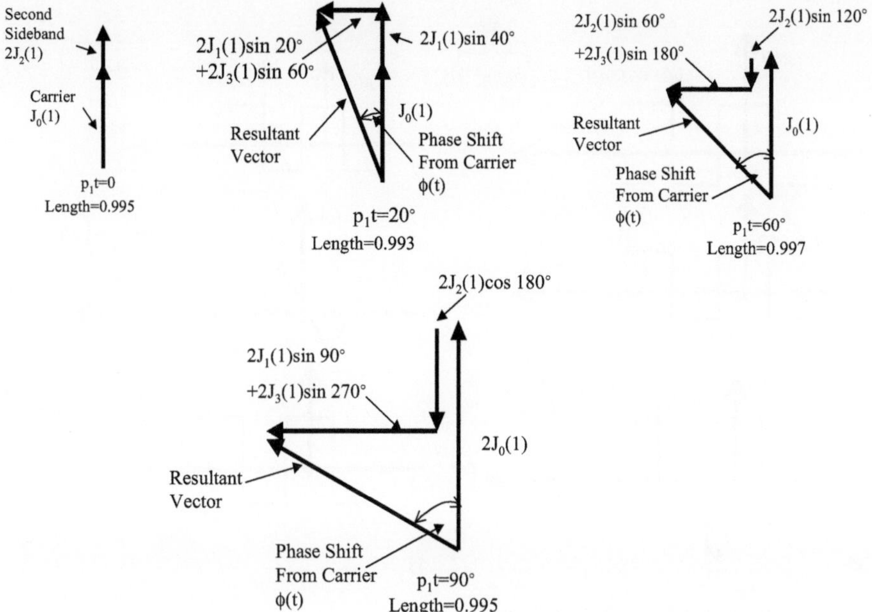

Fig. 1.13 Vector sum of composite wideband FM waveform at various times p_1t=0°, 20°, 60°, 90°

modulation function; and (2) It will also show that the amplitude of the envelope does remain constant.

Suppose we choose Mp=1. Then, from Table 1.1, we find that only the first three sidebands need be considered. Also, from this table we have as the values of interest: $J_0(1)$=0.765, $J_1(1)$=0.44, $J_2(1)$=0.115. Figure 1.13, shows the composite waveform with p_1t=0°, p_1t=20°, p_1t=60°, p_1t=90°.

1.4.5   Phase Modulation Spectrum

Phase and frequency modulation are sometimes lumped together and called *angular modulation*. However, some differences do exist, which merit investigation. These, we now consider.

In a phase modulated signal, the envelope is, again, held constant and the phase changed in a manner dictated by,

$$\varphi = \varphi_0\, mg(t) + \omega_0 t \tag{1.57}$$

Also, if we consider a time function of the form,

$$v(t) = A\cos\varphi \tag{1.58}$$

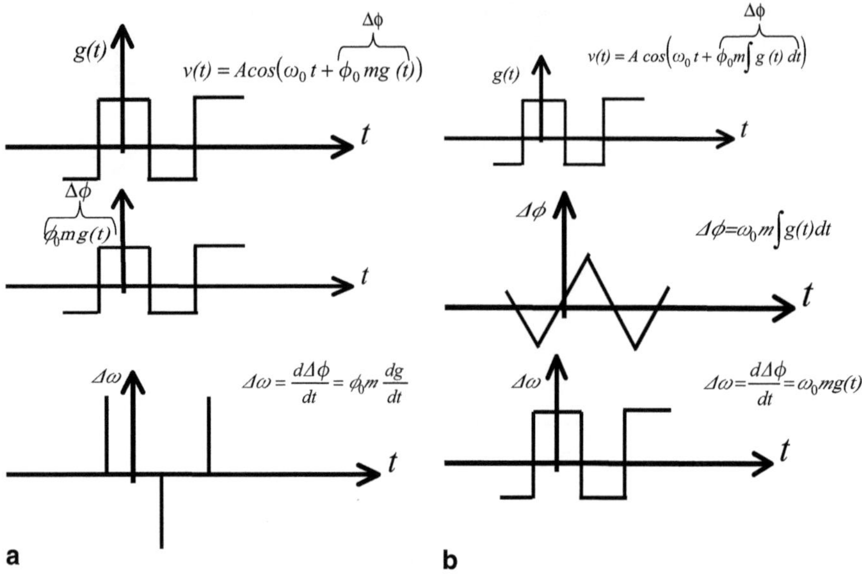

Fig. 1.14 Graphical description of PM and FM

the phase-modulated time function becomes,

$$v(t) = A\cos\left[\omega_0 t + \varphi_0 mg(t)\right] \tag{1.59}$$

The instantaneous frequency of the phase-modulated signal becomes, from (1.57),

$$\omega(t) = \frac{d\varphi}{dt} = \omega_0 + \varphi_0 m\frac{dg(t)}{dt} \tag{1.60}$$

If we consider a sinusoidal modulating function,

$$g(t) = \sin p_1 t \tag{1.61}$$

and now substitute (1.61) and (1.59), the result becomes,

$$v(t) = A\cos\left[\omega_0 t + \varphi_0 m \sin p_1 t\right] \tag{1.62}$$

A comparison of (1.62) and the corresponding expression for an FM function (1.25) shows that PM, like FM, produces an infinite number of sidebands. The relative magnitude of these sidebands can be determined by simple substitutions. Figure 1.14 illustrates the differences in both instantaneous frequency and phase produced when a carrier is first frequency and then phase modulated. In each case, the modulating function is a square wave [1].

1.5 Summary

In this chapter, we have presented an overview of wireless communication systems. We began by defining the fundamental wireless communications problem and motivating approaches to its solution. This was followed by a number of definitions, namely, those of carrier, baseband, and modulation. Then, upon introducing the simplified block diagram of a transmitter, a description of its typical building block elements, was presented. We then focused on the mathematical description and corresponding spectral properties of amplitude modulation (AM), frequency modulation (FM), and phase modulation (PM). This was then followed by a study of differences among modulation schemes, their bandwidth, and their vectorial representation, including the separate cases of narrow- and wide-band FM. The chapter was concluded with a discussion of phase modulation (PM) and its relationship to FM.

References

1. H. J. De Los Santos, Lecture Notes, "Communications Circuits," Course EE115D, Department of Electrical Engineering, University of California, Los Angeles (UCLA), Fall Quarter, 1990.
2. H. L. Krauss, C. W. Bostian, and F. H. Raab, *Solid State Radio Engineering*, John Wiley & Sons, Inc., 1980.

Chapter 2
Modulation and Detection

Abstract In this chapter, we present a number of topics surrounding modulation and demodulation. We begin by introducing system-level block diagrams of AM and FM/PM modulators and demodulators, and explaining their respective principles of operation. In particular, under the topic of AM Modulator/Demodulator, we introduce the full carrier modulator, the single sideband suppressed carrier modulator, the double sideband suppressed carrier modulator, the envelope detector, and the synchronous detector. Under the topic of FM and PM Modulator/Demodulator, we introduce the VCO as FM modulator, the indirect FM modulator, the PM modulator, the balanced discriminator FM demodulator, the quadrature FM detector, the PLL-based FM detector, the zero-crossing FM detector, and the PM demodulator. Then, under the topic of digital modulation, we introduce the concepts of Nyquist Limit, data rate, Shannon Limit, information capacity, and bandwidth efficiency, as well as specific modulation schemes, such as, binary modulation, amplitude-shift keying (BASK), frequency-shift keying (BFSK), and phase-shift keying (BPSK), differential binary phase-shift keying (DBPSK), quadrature phase-shift keying (QPSK), $\pi/4$ shifted QPSK, minimum shift keying (MSK), M-ary quadrature amplitude modulation (QAM), orthogonal frequency division multiplexing (OFDM), direct sequence spread spectrum (DS/SS), and frequency hopping spread spectrum (FH/SS). We also introduce the geometric representation of digital modulation schemes and the complex envelope form of a modulation signal.

2.1 AM Modulators

An AM modulator, which implements the equation,

$$v_{AM}(t) = a_0\left[1 + mg(t)\right]\cos\omega_0 t \tag{2.1}$$

may be represented as shown in Fig. 2.1. It is seen that the magnitude of the modulation index, m, if too large may cause distortion of the resulting modulated signal.

The corresponding frequency spectrum for the AM signal is given by [1],

$$F_M(\omega) = \frac{a_0}{2}[2\pi\delta(\omega - \omega_0) + 2\pi\delta(\omega + \omega_0) + F_B(\omega - \omega_0) + F_B(\omega + \omega_0)] \tag{2.2}$$

which is depicted in Fig. 2.2. As can be seen, due to symmetry with respect to the origin, the same information is carried in both the upper (centered at $+\omega_0$) and lower

H. J. De Los Santos et al., *Radio Systems Engineering,*
DOI 10.1007/978-3-319-07326-2_2, © The Authors 2015

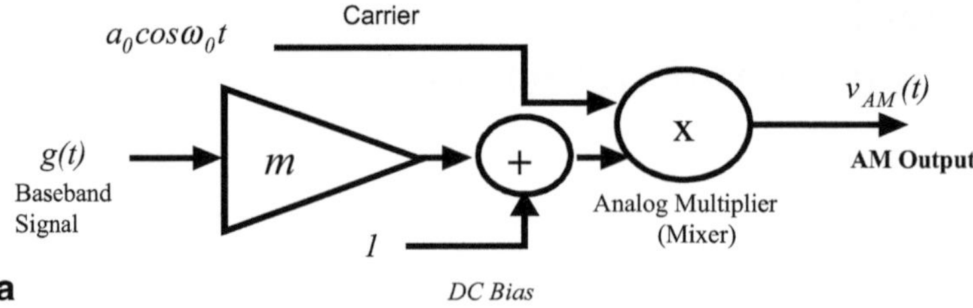

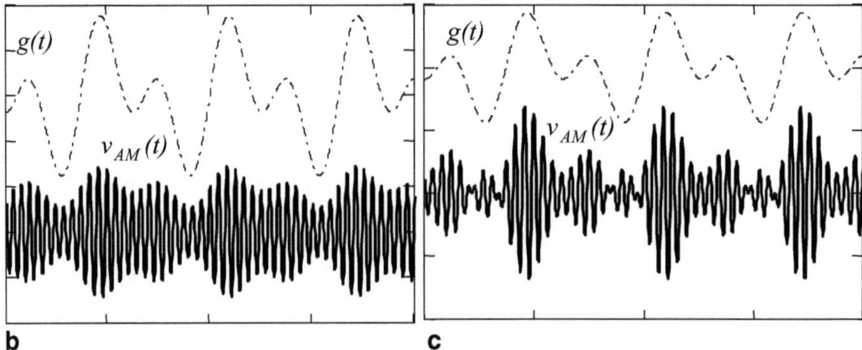

Fig. 2.1 a Depiction of system-level block diagram of AM modulator. **b** Depiction of modulating function, $g(t)$, and resulting modulated function, $v_{AM}(t)$ for the case of a correctly modulated AM signal, in which $(1+mg(t)) > 0$. **c** Depection of modulating function, $g(t)$, and resulting modulated function, $v_{AM}(t)$ for the case of an incorrectly modulated AM signal, in which $(1+mg(t)<0)$. In this case the modulation index, m, is too large and it causes distortions in the amplitude envelope. *After* [1]

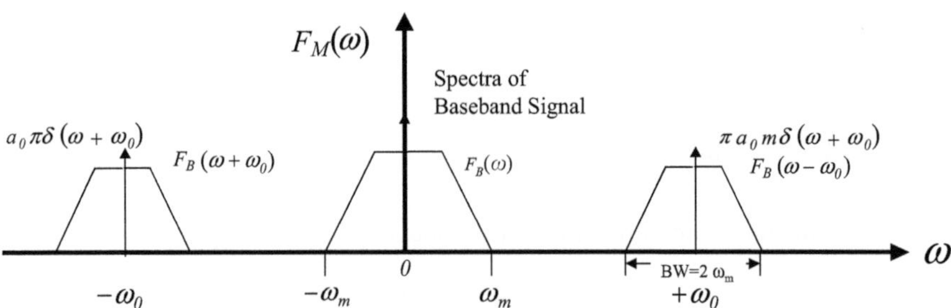

Fig. 2.2 Spectra of AM signal. The spectra of the baseband signal, centered at zero frequency, possess a maximum frequency ω_m

(centered at $-\omega_0$) sidebands. Therefore, not only one of them is redundant, but the transmitted power is split between the two sidebands. Finally, the modulated signal, centered on ω_0, occupies twice ω_m, i.e.,

$$BW_{AM} = 2\omega_m \qquad (2.3)$$

The excess, unnecessary, power and bandwidth accompanying the AM signal represented by (2.2), motivated another scheme to eliminate one of the sidebands, namely, *single sideband* AM or SSB modulation.

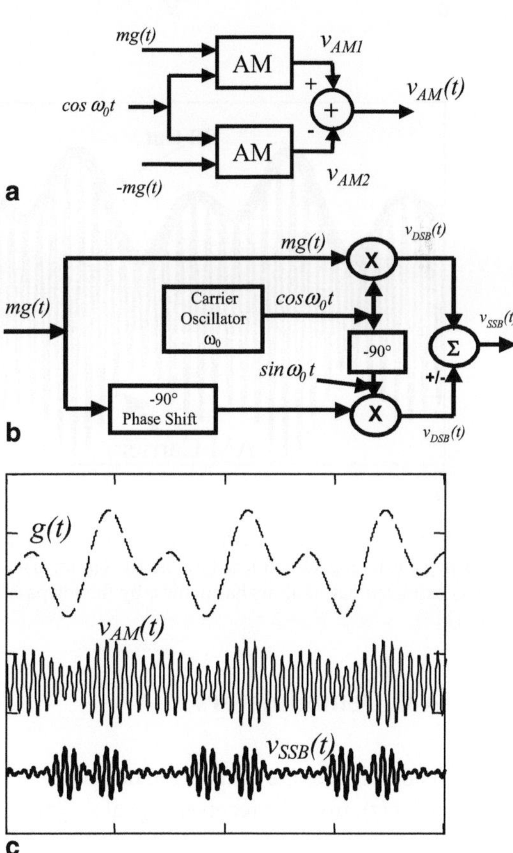

Fig. 2.3 Depiction of filtering method for producing a single sideband (SSB) AM signal. *After* [1]

Fig. 2.4 **a**, **b** Block diagram depicting the balanced modulator method to create a single sideband AM signal. **c** Modulation signal, $g(t)$, resulting modulated signal with full carrier, $v_{AM}(t)$, and resulting modulated signal with suppressed carrier, $v_{SSB}(t)$. *After* [1]

2.1.1   Methods to Create Single Sideband (SSB) AM Signals

Two main methods to create an SSB signal are presented, namely, the filtering method and the balanced modulation method. In the filtering method, whose block diagram is presented in Fig. 2.3, one of the sidebands is eliminated by direct filtering.

In the balanced modulator method, the carrier is modulated with both $+g(t)$, and $-g(t)$, and then subtracted, as depicted in Fig. 2.4a. This operation is implemented by the block diagram shown in Fig. 2.4b (for the case in which there is no DC added to $g(t)$), and results in the waveforms $v_{SSB}(t)$ shown in Fig. 2.4c [1].

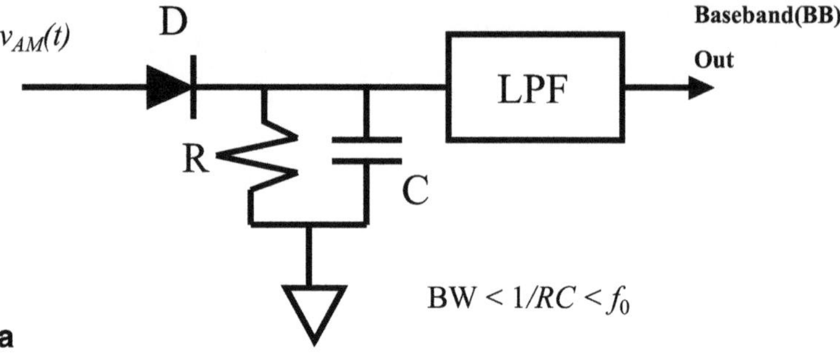

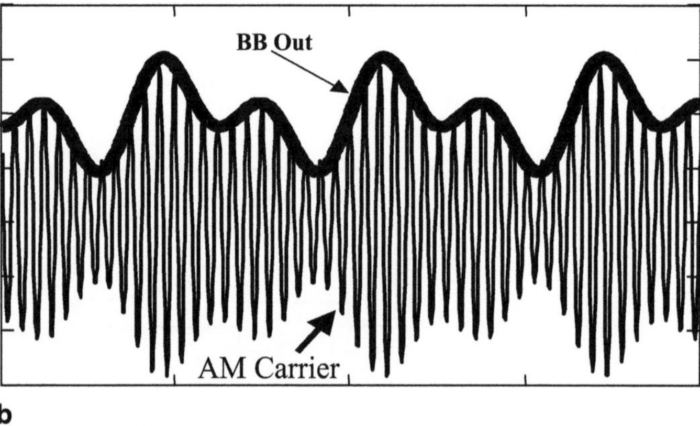

Fig. 2.5 **a** Envelope detector circuit plus lowpass filter. **b** Modulated AM signal, $v_{AM}(t)$. The high frequency carrier signal is filtered by the RC circuit. The modulating (baseband) *envelope* signal, $g(t)$, is extracted out of all its harmonics, by the lowpass filter and delivered to the output as BB. *After* [1]

2.2   AM Demodulators

A variety of schemes are available to demodulate, i.e., to extract, the information-bearing signal, $g(t)$, from the received signal $v_{AM}(t)$ or $V_{SSB}(t)$; these are discussed next.

2.2.1   Envelope Detector

The *Envelope Detector* followed by a lowpass filter is perhaps the most fundamental AM demodulation system, Fig. 2.5. In this circuit, the received input signal is rectified by a diode, D, and passed by an RC filter which, with an inverse time constant greater than the modulating signal bandwidth, but smaller than the carrier frequency, is not responsive to the high-frequency carrier, but only to the modulating signal, the signal envelope and its harmonics. The harmonics are then attenuated, by a lowpass filter which selects the modulating signal [1].

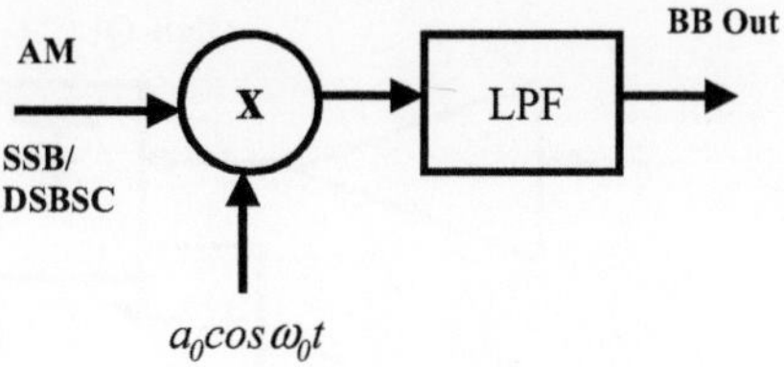

Fig. 2.6 Product detector plus lowpass filter. When the locally generated carrier is synchronized to the transmitter, the circuit is called *synchronous detector. After* [1]

2.2.2   Product Detector

In the product detector, the incoming signal is multiplied by a locally generated carrier, and the resulting signal is passed through a lowpass filter, Fig. 2.6 [1]. In particular, the product of the incoming signal with the locally generated carrier yields,

$$K\left[1+mg(t)\right]\cos^2 \omega_0 t = K\left[1+mg(t)\right]\cdot\frac{1}{2}(1+\cos\ 2\omega_0 t)$$

$$\rightarrow K\left[1+mg(t)\right]\cdot\frac{1}{2}+K\left[1+mg(t)\right]\cdot\frac{1}{2}\cdot\cos\ 2\omega_0 t \qquad (2.4)$$

which contains a first term whose spectral content is that of the baseband signal, and a second term whose spectrum is centered at twice the carrier frequency, i.e., the second term yields a double-sideband suppressed carrier (DSBC) signal. Therefore, at the output of the lowpass filter, only the baseband signal remains.

2.3   FM Modulators

Two FM demodulators will be presented, namely, the direct and the *indirect* methods.

2.3.1   Direct FM Modulator

The direct FM modulator uses a voltage controlled oscillator, Fig. 2.7.

In this circuit, the VCO has a free-running frequency, f_0, given by,

$$f_0 = \frac{1}{2\pi\sqrt{L_0 C_0}} \qquad (2.5)$$

This frequency is modulated by changing the value of frequency-determining capacitors (varactors) through the application of the baseband signal. In particular, the VCO output frequency becomes a time-dependent function given by,

$$f(t) = f_0 + kg(t) \qquad (2.6)$$

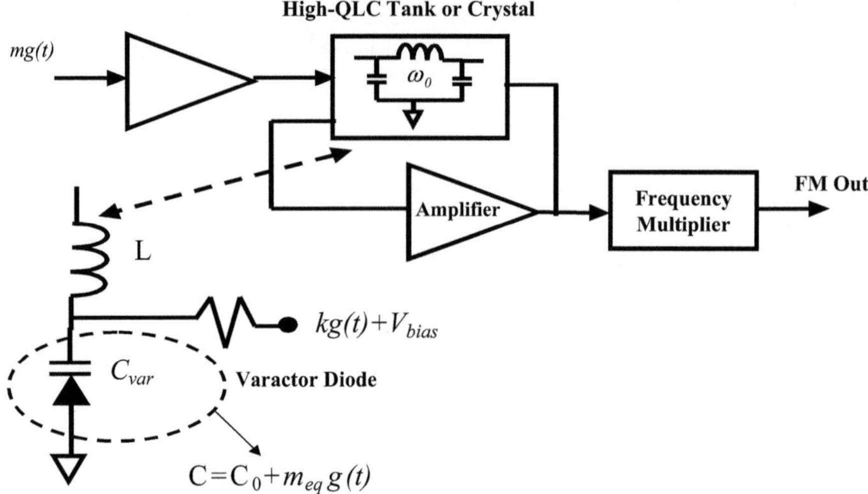

Fig. 2.7 Voltage-controlled oscillator (VCO) as FM modulator. *After* [1]

Fig. 2.8 Baseband signal (*top*) and frequency modulated carrier (*bottom*). *After* [1]

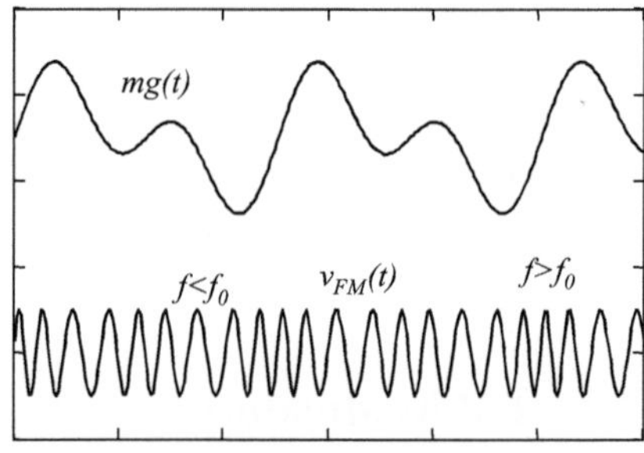

As the amplitude of the baseband signal increases and decreases, the carrier frequency follows, Fig. 2.8 [1].

2.3.2   Indirect FM Modulator

Rather than modulating the varactor voltage of a VCO, in the indirect FM method the modulating signal is first integrated, then used to phase-modulate a crystal-controlled carrier frequency. The signal is then multiplied for producing a wideband FM signal. The block diagram of an indirect FM modulator is shown in Fig. 2.9.

2.3.3   PM Modulator

To effect phase modulation, the constant carrier signal is generated independently, and then passed through an LC circuit whose capacitance is modulated by the

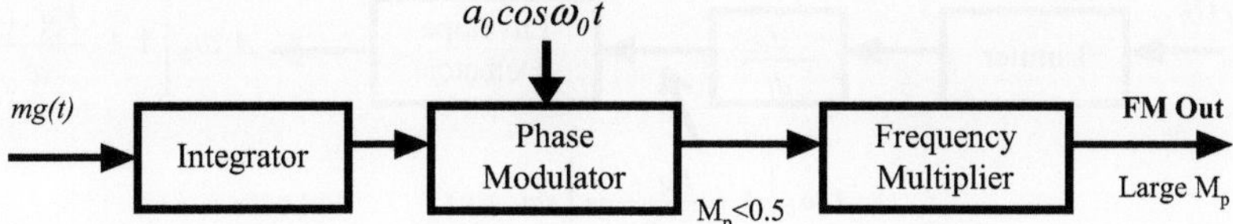

Fig. 2.9 Indirect FM modulator. *After* [1]

Fig. 2.10 Phase modulator circuit. *After* [1]

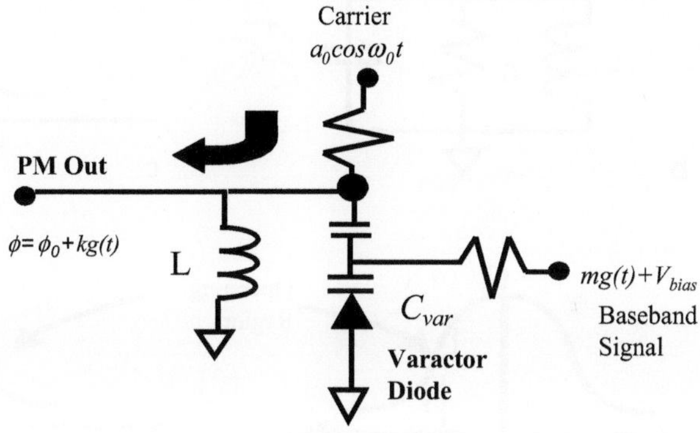

baseband signal, Fig. 2.10. For small varactor voltage swings, the LC circuit is linear and small phase shifts (a few degrees) are obtained. Larger deviations may be obtained using a frequency multiplier. The time-domain waveform of PM is similar to that of FM, and can only be distinguished from FM from knowledge of the source.

2.4   FM Demodulators

The various methods for effecting FM demodulation may be surmised from examining the FM waveform, namely [1],

$$v_{FM}(t) = A\cos\varphi(t) = A\cos\left[\omega_0 t + K\int g(t)dt\right] \tag{2.7}$$

In the first place, differentiating the instantaneous frequency,

$$\omega_i(t) = \frac{d\varphi(t)}{dt} = \omega_0 + Kg(t) \tag{2.8}$$

it is seen that we get the desired output g(t), if we cancel ω_0.

On the other hand, differentiating $v_{FM}(t)$, we obtain:

$$\frac{dv_{FM}(t)}{dt} = -A\sin\varphi(t)\frac{d\varphi}{dt}$$
$$= -A\left[\omega_0 + Kg(t)\right]\sin\varphi(t) \tag{2.9}$$

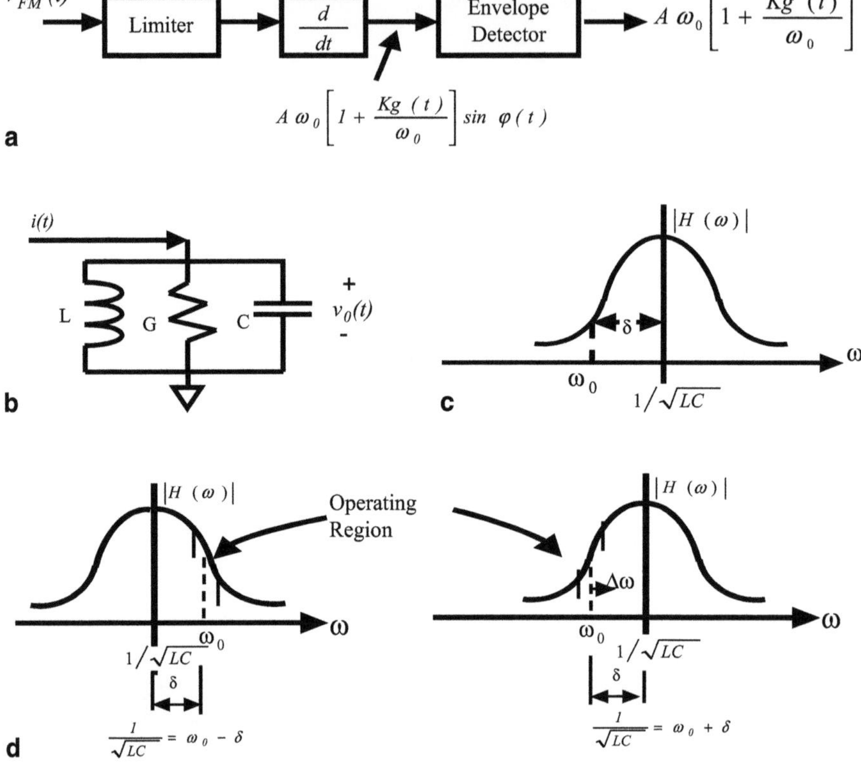

Fig. 2.11 a FM detector block diagram. **b** Circuit implementing differentiation (d/dt) block. **c** Transfer function of GLC circuit in **b**. **d** Relation of frequency deviation to operating region.

which looks like the an AM signal with envelope given by,

$$A\left[\omega_0 + Kg(t)\right] = A\omega_0\left[1 + \frac{Kg(t)}{\omega_0}\right] \tag{2.10}$$

Therefore, we may obtain $g(t)$ by the envelope-detecting operation $dv_{FM}(t)/dt$.

2.4.1 FM Balanced Demodulator

The *balanced demodulator* implements FM demodulation as shown in Fig. 2.11a. In this block diagram, the *limiter* is inserted, prior to differentiation, to ensure that no terms coming from A result when the derivative is taken (otherwise these variations in A may obscure the true amplitude variations due to $g(t)$) [2].

Then, middle block in Fig. 2.11a, differentiation, $dv_{FM}(t)/dt$, is effected, which results in a signal whose amplitude is modulated by $g(t)$; this is like an AM signal with high-frequency carrier $\phi(t)$. The differentiation is implemented by a parallel GLC circuit, Fig. 2.11b, with transfer function,

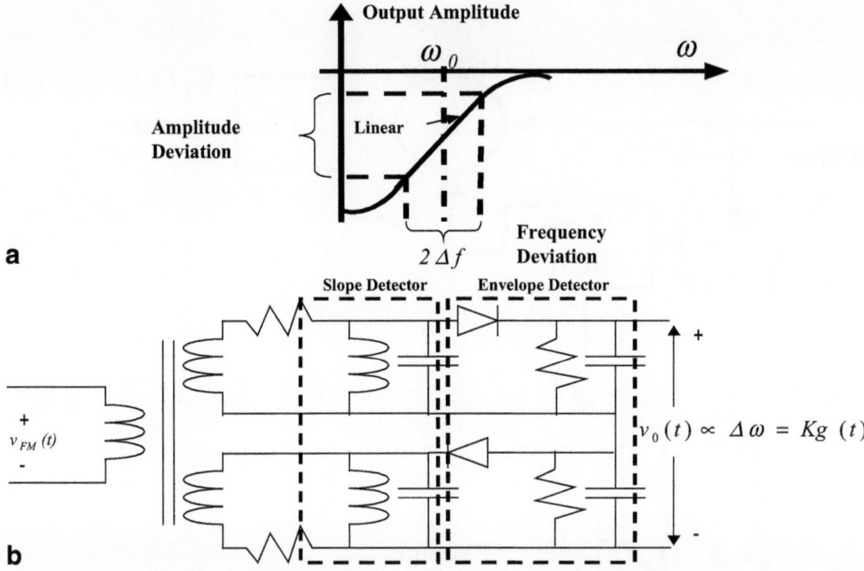

Fig. 2.12 a Characteristic of differentiation circuit, where a frequency deviation is translated into an amplitude deviation. **b** Balanced FM demodulator circuit differentiates and then detects FM signal, extracting baseband signal

$$|H(\omega)| = \left| \frac{V_0(\omega)}{I(\omega)} \right| = \frac{1/G}{\sqrt{1 + (\omega C - 1/\omega L)^2/G^2}} \tag{2.11}$$

The circuit has a damping $\alpha = G/2C$, and resonance frequency $1/\sqrt{LC}$, which is a distance δ from the carrier frequency, ω_0. If the carrier frequency, ω_0, is set at the middle of the positive slope of the transfer curve, Fig. 2.11d, then it will deviate a distance $\Delta\omega$, so that $\dfrac{1}{\sqrt{LC}} = \omega_0 + \delta$ on this operating region, and $\dfrac{1}{\sqrt{LC}} = \omega_0 - \delta$ on negative slope operating region. Then, assuming $\Delta\omega << \omega_0$, $\delta << \alpha$, and $\delta << \omega_0$, it can be shown [2] that the difference between H evaluated at these points is given by,

$$|H|_{\omega_0 + \delta} - |H|_{\omega_0 - \delta} = \frac{2\delta}{G\alpha^2} \Delta\omega \tag{2.12}$$

It will be recalled that, differentiation of a function of time with respect to time, in the time domain, is proportional to multiplication by frequency in the frequency domain, i.e., $d/dt \rightarrow j\omega$. Therefore, (2.12) is the frequency domain representation of differentiation [2].

Within a small deviation about the *high* carrier frequency ω_0, the slope of the transfer curve, *H*, is linear, Fig. 2.12a. While this differentiated signal is now an

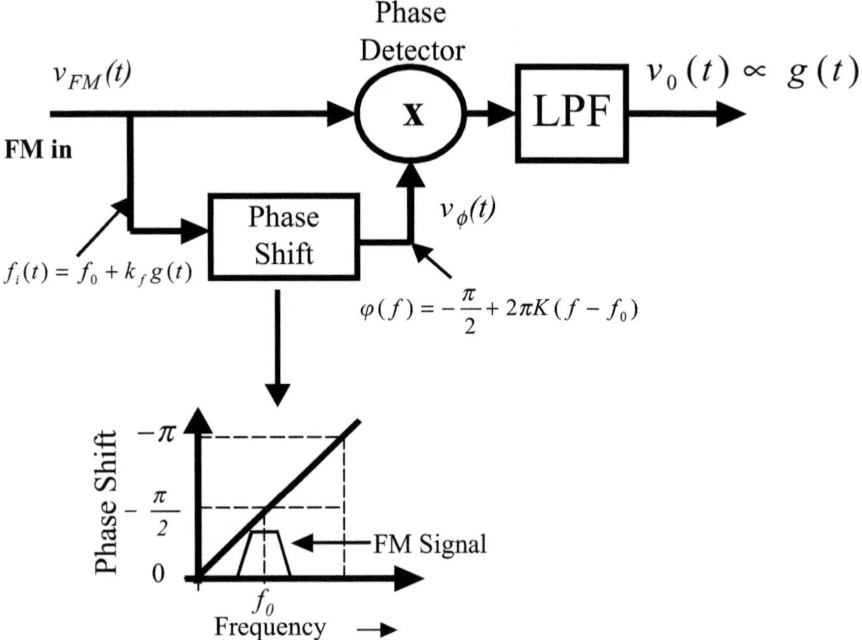

Fig. 2.13 FM demodulation by phase detector. *After* [1]

AM signal with its amplitude dictated by $g(t)$, but still at a high frequency, it must be passed by an envelope detector to remove the high-frequency carrier and finally extract $g(t)$. FM demodulation is, thus, completed by adding an envelope detector, Fig. 2.12b [1, 2], producing an output voltage given by,

$$v_0(t) \propto \Delta\omega = Kg(t) \tag{2.13}$$

2.4.2 *FM Quadrature (Detector) Demodulator*

Examination of an FM signal reveals that its phase is shifted by an amount proportional to its instantaneous frequency (2.14).

$$v_{FM}(t) = A\cos(\omega_0 t + \varphi(t)) \tag{2.14}$$

This fact is exploited by the product detector, Fig. 2.13 [1], to detect the phase difference between an incoming FM signal and the signal at the output of a phase shift network (2.15),

$$v_\varphi(t) = A'\cos\left[2\pi f_0 t + 2\pi k_f \int g(t')dt' + \varphi(f_i(t))\right] \tag{2.15}$$

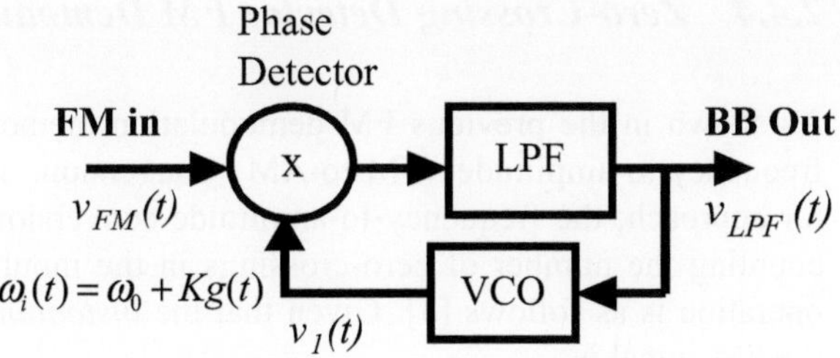

Fig. 2.14 Phase-locked loop as FM demodulator. *After* [1]

Since the phase shift introduced by the phase shift network is proportional to the instantaneous frequency of the input FM signal, the output voltage of the phase detector, after removing the high-frequency components by a lowpass filter, will also be proportional to the instantaneous frequency of the input FM signal, thus reproducing the original modulating signal, (2.16).

$$v_0(t) = A'^2 \cos(\phi(f_i(t))) = A'^2 \cos\left(-\frac{\pi}{2} + 2\pi K(f_i(t) - f_0)\right)$$
$$= A'^2 \sin\left[2\pi K k_f g(t)\right] \propto g(t) \tag{2.16}$$

The latter conclusions is clearly seen when one considers the series expansion of the sine function, which would contain the fundamental, which is proportional to $g(t)$, the third power of $g(t)$, etc. Thus, only the fundamental term, proportional to $g(t)$, would survive the lowpass filter, hence, (2.16).

2.4.3   Phase-Locked Loop (PLL)-Based FM Demodulator

The PLL, Fig. 2.14 [1], is a closed loop control system that tracks the variations in the received signal phase and frequency (2.17).

$$v_{FM}(t) = A\cos(\omega_0 + \Delta\omega)t \tag{2.17}$$

In this scheme, the VCO output,

$$v_1(t) = A_1 \cos(\omega_0 t + \varphi_1(t))t \tag{2.18}$$

where,

$$\omega_1 = \omega_0 + \frac{d\varphi_1(t)}{dt} \tag{2.19}$$

is compared with the input signal using a phase comparator (detector). The phase comparator produces an output voltage proportional to phase difference. The phase difference, in turn, is low-pass-filtered, producing $V_{LPF}(t)$, and fed back to the VCO to control its output frequency. The control of the VCO, $V_{LPF}(t)$, is itself the demodulated FM signal, $g(t)$ [1].

2.4.4　Zero-Crossing Detector FM Demodulator

As shown in the previous FM demodulation methods, they typically involve a frequency-to-amplitude (FM-to-AM) conversion. In the zero-crossing detector approach, the frequency-to-amplitude conversion operation is performed by counting the number of zero-crossings in the input FM signal; its principle of operation is as follows [3]. Given that the *instantaneous frequency deviation* in the FM signal is,

$$\omega_i = \omega_0 + \frac{d\varphi_i(t)}{dt} \tag{2.20}$$

then,

$$\int_{t_1}^{t_2} d\varphi(t) = \int_{t_1}^{t_2} 2\pi f_i(t)dt \tag{2.21}$$

from where, upon integration between adjacent zero-crossing times t_1 and t_2, where $t_2 > t_1$, one obtains,

$$\varphi(t_2) - \varphi(t_1) = \int_{t_1}^{t_2} [2\pi f_0 + mg(t)]dt \tag{2.22}$$

But, also, because t_1 and t_2 are adjacent zero crossings, we must have,

$$\varphi(t_2) - \varphi(t_1) = \pi \tag{2.23}$$

or

$$\left[2\pi f_0 + mg(t)\right]\cdot\left(t_2 - t_1\right) \approx \pi \tag{2.24}$$

which, using $2\pi f_i(t)$, may be written as,

$$\pi \approx 2\pi f_i(t) \cdot (t_2 - t_1) \tag{2.25}$$

from where we obtain,

$$f_i(t) = \frac{1}{2\pi (t_2 - t_1)} \tag{2.26}$$

Now, since,

$$f_i(t) = f_0 + \frac{mg(t)}{2\pi} \tag{2.27}$$

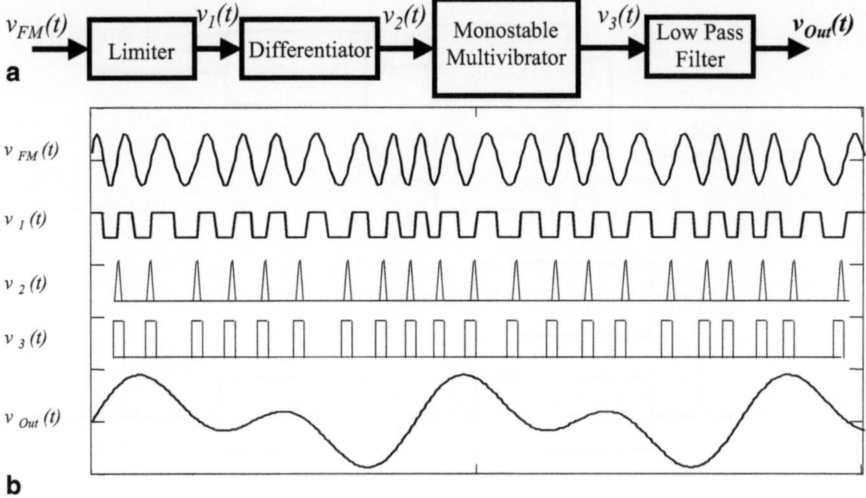

Fig. 2.15 **a** Block diagram of zero-crossing detector for FM demodulation. **b** Waveforms throughout the demodulator

Fig. 2.16 **a** FM demodulator transfer curve. **b** PM demodulator transfer curve

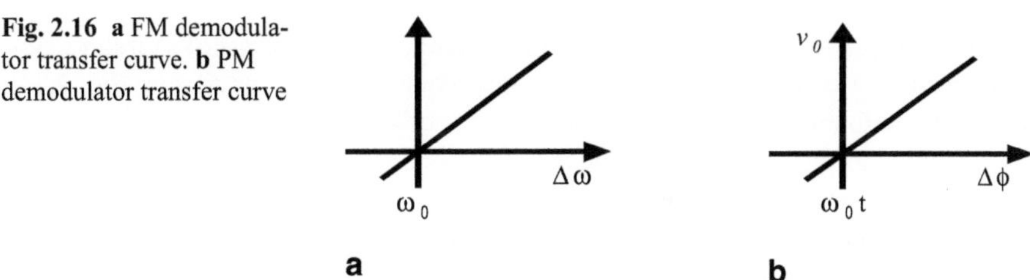

$g(t)$ may be found by measuring the spacing between zero-crossings in the interval t_2-t_1.

The implementation of this method is depicted by the block diagram in Fig. 2.15 [3].

Following Fig. 2.15 (from top to bottom), the input FM signal, $v_{FM}(t)$, is applied to a limiter which, upon clipping the signal renders a rectangular pulse train, $v_1(t)$, reflecting the FM time variation. This FM pulse train, in turn, is applied to a differentiator, producing sharp pulses $v_2(t)$. These sharp pulses, coming out of the differentiator, are then employed as triggers to a monostable multivibrator (a "one shot"), which produces a pulse train $v_3(t)$, such that its average duration is proportional to the desired (modulating) baseband signal. Upon passing this pulse train by a lowpass filter, its average, a slowly varying DC component, $v_{Out}(t)$, is extracted; this is the desired demodulated signal, $g(t)$ [3].

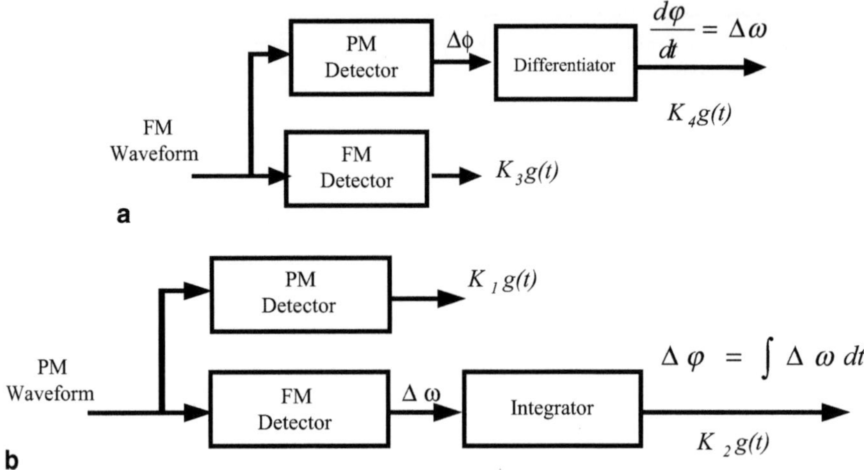

Fig. 2.17 **a** FM demodulator. **b** PM demodulator

2.5 PM Demodulators

As described in the last few sections, an FM demodulator consists of a circuit whose output voltage is proportional to the instantaneous *frequency* deviation, Fig. 2.16a. A PM demodulator, on the other hand, consists of a circuit whose output voltage is proportional to the instantaneous *phase* deviation, Fig. 2.16b [4].

Since the phase is the integral of the frequency, one should be able to perform the demodulation function for either type of demodulation, FM or PM, with a single detector, FM or PM, and an integrator or differentiator; this is shown in Fig. 2.17 [4].

2.6 Digital Modulation

In digital modulation, the carrier amplitude, frequency, or phase are made to vary (i.e., are modulated) according to a *digital representation* of the baseband signal. The first step in digital modulation, therefore, is to convert the baseband signal into a digital stream of bits. This entails sampling the baseband signal at a frequency fs, and representing each sample by N quantization bits, Fig. 2.18 [1].

As is known, to enable recovery of the original signal from the sampled signal, the original signal must be sampled at a frequency greater than twice its maximum frequency, B_m. This is the so-called *Nyquist sampling rate limit*. If the original signal is sampled every $t_S = 1/f_S$ seconds, and every sample contains N bits, then the rate of creation of samples is $N/t_S = f_S N$ bits/s. This is the sampled data rate, R. It is said, then, that the original signal is represented by a *data rate* of R bits/s [1].

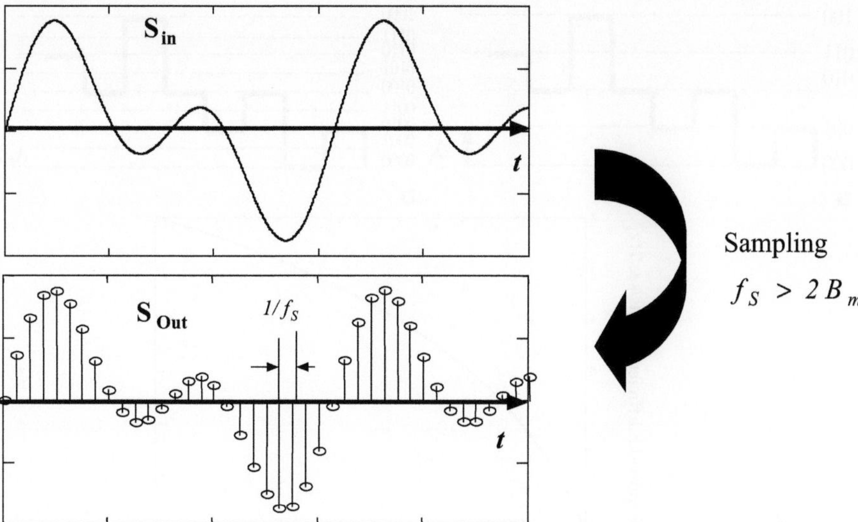

Fig. 2.18 Digitization of input baseband signal, S_{in}, via sampling at time intervals $1/f_S$, to produce a sampled output representation, S_{Out}, where each sampling point is in turn represented by N bits. *After* [1]

The data rate, R, utilized in transmitting a given signal, may not be increased arbitrarily by increasing f_S, but it is limited by the bandwidth of the channel through which the signal travels. This capacity, in particular, is given by [1],

$$C = 2BLog_2M \quad [bits/\sec] \tag{2.28}$$

where B is the channel bandwidth and M is the number of bits utilized to represent a signal level being transmitted. For example, if we use two bits to represent a signal level, then M=2. If three bits, M=3, see Fig. 2.19. C is known as the *Shannon limit*. and represents the maximum bit rate allowed by a channel. It is evident from Fig. 2.19 that, while the resolution with which a given signal level may be represented increases with the number of bits utilized, the fact that the step-size decreases as M increases results in an increase in the likelihood that noise might "push" one level into the next, thus contributing to an error, i.e., it is easier for noise to confuse the true quantized level. A standardized parameter that may be used to compare channel capacity is the so-called *bandwitdh efficieny*, which is given by [1],

$$\frac{C}{B} = 2Log_2M \quad [bits/\sec/Hz] \tag{2.29}$$

A plot of bandwidth efficiency versus M is shown in Fig. 2.19.

A standardized parameter that may be used to compare channels is the so-called *bandwitdh efficieny*, which is given by,

$$\frac{C}{B} = 2Log_2M \quad [bits/\sec/Hz] \tag{2.30}$$

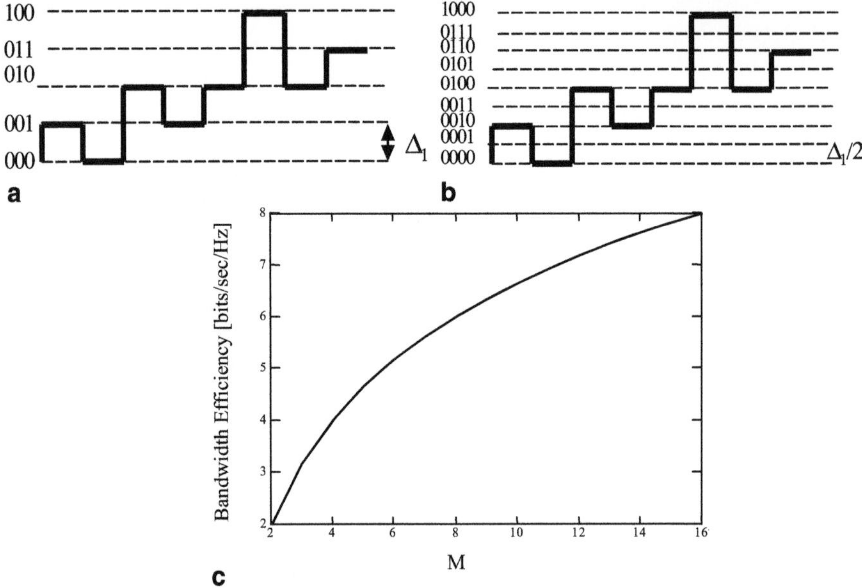

Fig. 2.19 Signal quantization. **a** Signal levels represented by 3 bits, i.e., M=3 bits. **b** Signal levels represented by 4 bits, i.e., M=4 bits. **c** Bandwidth efficiency versus number of bits representing a signal level. *After* [1]

A plot of bandwidth efficiency versus M is shown in Fig. For M=3, the maximum bandwidth efficiency would be 3.17.

A measure of channel capacity, that doesn't require that M be specified, is in terms of the signal-to-noise ratio (SNR). The SNR is defined as,

$$SNR = \frac{E_b R}{N_0 B} \tag{2.31}$$

where E_b is the average energy per bit, and N_0 is the noise power spectral density. In practice, $C < R$, therefore, one obtains,

$$\frac{C}{B} = \text{Log}_2\left(1 + \frac{E_b}{N_0} \cdot \frac{R}{B}\right) \tag{2.32}$$

What this formula indicates is that, for a given noise level, N_0, channel capacity is a function of the average energy per bit, E_b, expended [1–3].

2.6.1 *Binary Modulation—Amplitude-Shift Keying (ASK)*

When the baseband signal is quantized into two levels, N=2, it is represented by a binary stream of 1's and 0's. Binary modulation of the carrier amplitude is called

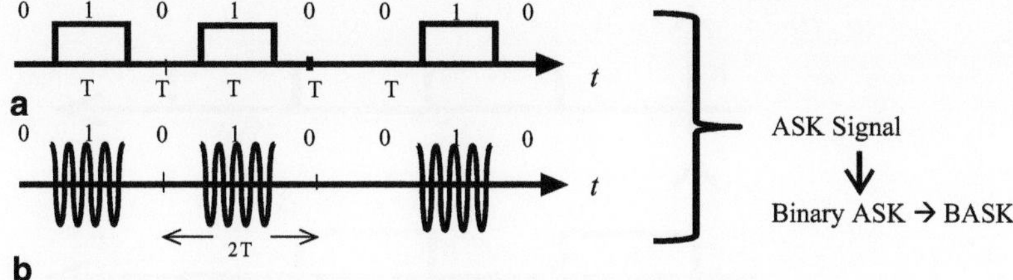

Fig. 2.20 Amplitude-Shift-Keying. **a** Binary signal. **b** Modulated signal. *After* [7]

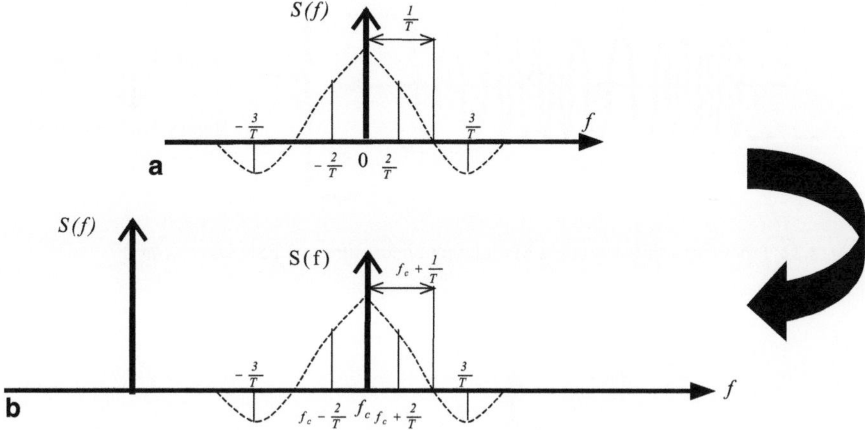

Fig. 2.21 **a** Spectrum of baseband signal. **b** Spectrum of ASK Wave (positive frequencies only). *After* [7]

ON-OFF Keying (OOK) or Amplitude-Shift-Keying (ASK), Fig. 2.20, whose corresponding spectrum is given in Fig. 2.21.

2.6.2 Binary Modulation—Frequency-Shift Keying (FSK)

Binary modulation of the carrier frequency is called Frequency-Shift Keying (FSK). In this scheme, as depicted in Fig. 2.22, the transmitted frequency is dictated by a binary sequence of 1's and 0's, according to which frequencies f_1 and f_0 are transmitted over intervals of time T; the corresponding spectrum is as depicted in Fig. 2.23.

2.6.3 Binary Modulation—Phase-Shift Keying (PSK)

Binary modulation of the carrier phase is called Phase-Shift Keying (PSK). This is illustrated in Fig. 2.24. In this scheme, the phase of the carrier is shifted 180° in

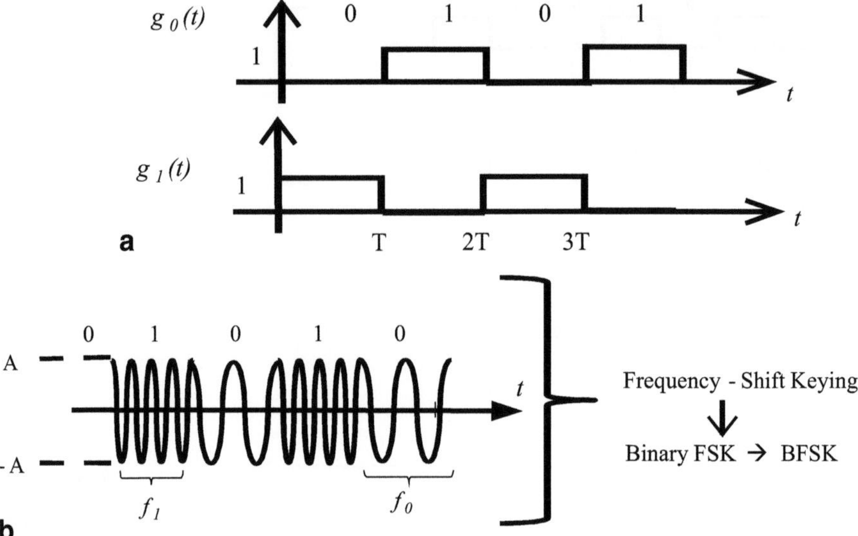

Fig. 2.22 Frequency-shift-keying. **a** Modulating binary signal. **b** FSK wave. *After* [7]

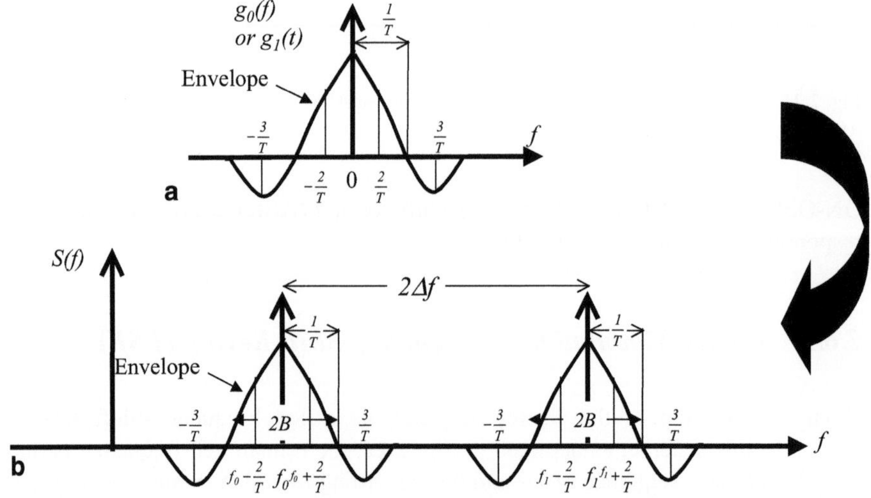

Fig. 2.23 a Spectrum of modulating baseband signal. **b** Spectrum of FSK Wave (positive frequencies only). *After* [7]

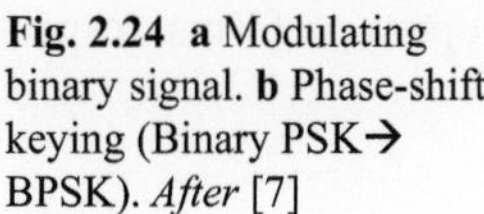

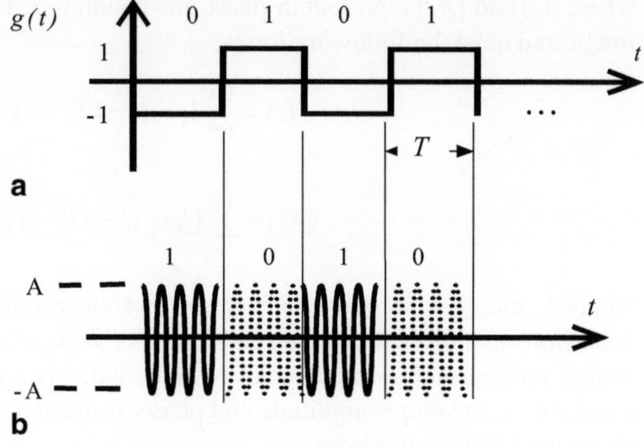

Fig. 2.24 a Modulating binary signal. b Phase-shift keying (Binary PSK→ BPSK). *After* [7]

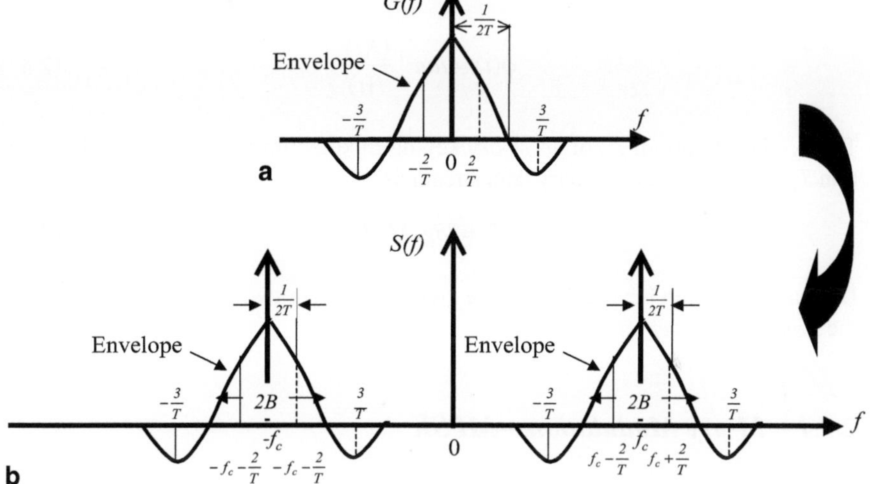

Fig. 2.25 a Spectrum of modulating baseband signal. b Spectrum of PSK wave. *After* [7]

accord with whether the modulating signal is a 1 or a 0. The pertinent spectrum is shown in Fig. 2.25.

2.7   Complex Envelope Form of Modulation Signals

Waveforms that are employed for digital modulation may be expressed in complex envelope form. [3, 5] i.e.

$$M(t) = I(t) + j Q(t) = A(t)e^{j\varphi(t)} \tag{2.33}$$

where $I(t)$ and $Q(t)$ represent in-phase and quadrature envelope waveforms, respectively, and have the following forms,

$$I(t) = \sum_k I_k p_I(t - kT_s - \tau) \tag{2.34}$$

$$Q(t) = \sum_k Q_k p_Q(t - kT_s - \tau) \tag{2.35}$$

where I_k and Q_k represent sequences of discrete variables mapped from the baseband (information) data with a symbol rate of $1/Ts$, $pI(t)$ and $pQ(t)$ represent finite energy pulses, such as rectangular or Gaussian, τ is a possible delay, and $A(t$ and $\varphi(t))$ are the envelope amplitude and phase, respectively [3, 5]. The amplitude and phase in (2.33) are given by,

$$A(t) = \sqrt{I^2(t) + Q^2(t)} \tag{2.36}$$

$$\phi(t) = tan^{-1} \frac{Q(t)}{I(t)} \tag{2.37}$$

The symbol duration, T_s, of an *M-ary* keying modulation is related to the bit duration T_b of the originally bynary data stream as,

$$T_s = log_2 M \cdot T_b \tag{2.38}$$

Digital modulation beyond binary (i.e., with more than two levels) usually exploits the *I and Q representation*.

2.7.1   M-ary Modulation—MPSK

The representation of these modulation schemes is better visualized via the concept of constellations. We begin this subject by introducing the *geometric* representation of modulation signals.

If there are M possible signals, the modulation signal set S can be represented as [5]:

$$S = \{s_1(t), s_2(t), ..., s_M(t)\} \tag{2.39}$$

While for binary modulation schemes, a binary information bit is mapped directly to a signal and S will contain only two signals, for higher-level modulation schemes (*M-ary* keying) the signal set will contain more than two signals; each signal (or *symbol*) will represent more than a single bit of information.

With a signal set of size M, it is possible to transmit a maximum of $log_2 M$ bits of information per symbol. Since the elements of S may be viewed as points in a vector space, the representations are called *constellations*. When the baseband signal is quantized into M levels, $N = M$, it is represented by an *M-ary* stream of 1's and 0's. Now, from the fact that any finite set of physically realizable waveforms in a vector space can be expressed as a linear combination of N orthonormal waveforms, which form the basis of that vector space, representing the modulation signals on a vector space, requires finding a set of signals that form a basis for that vector space. Once a basis is determined, any point in that vector space can be represented as a linear combination of the basis signals [3, 5],

$$\left\{ \varphi_j(t) \big| j = 1, 2, ..., N \right\} \tag{2.40}$$

such that,

$$s_i(t) = \sum_{j=1}^{N} s_{ij} \phi_j(t) \tag{2.41}$$

The basis signals are orthogonal to one another such that,

$$\int_{-\infty}^{\infty} \phi_i(t)\phi_j(t)dt = 0 \quad i \neq j \tag{2.42}$$

and each of the basis signals is normalized to have unit energy, i.e.,

$$E = \int_{-\infty}^{\infty} \varphi_i^{\,2}(t)dt = 1 \tag{2.43}$$

2.7.2   Binary Phase Shift Keying Modulation—BPSK

In this digital modulation scheme, the phase of a constant envelope carrier is switched between two signals m_1 (bit 1) and m_2 (bit 0), the two phases being separated by 180°. The BPSK signal is [3],

$$s_{BPSK}(t) = \sqrt{\frac{2E_b}{T_b}} \cos(2\pi f_c t + \theta_c) \quad 0 \leq t \leq T_b \quad (\textit{binary } 1) \tag{2.44}$$

or

$$s_{BPSK}(t) = -\sqrt{\frac{2E_b}{T_b}} \cos(2\pi f_c t + \theta_c) \quad 0 \leq t \leq T_b \quad (\textit{binary } 0) \tag{2.45}$$

which may be written as,

$$s_{BPSK}(t) = m(t)\sqrt{\frac{2E_b}{T_b}} \cos(2\pi f_c t + \theta_c) \tag{2.46}$$

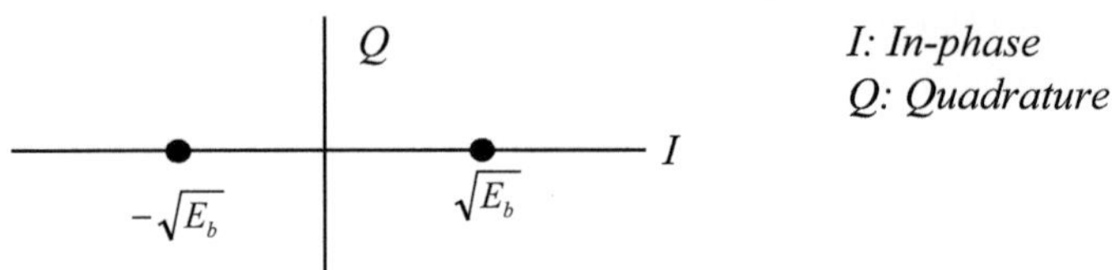

Fig. 2.26 BPSK constellation diagram

This is equivalent to double sideband suppressed carrier modulated waveform, where $m(t)$ is the modulating signal and $cos(2\pi f_c t)$ is the carrier; it can be generated using a balance modulator.

Assuming a rectangular pulse ($p(t) = rect((t\text{-}T_b/2)T_b)$, for BPSK signals $s_1(t)$ and $s_2(t)$ given by,

$$s_1(t) = \sqrt{\frac{2E_b}{T_b}}\ cos\,(2\pi f_c t) \quad 0 \leq t \leq T_b \tag{2.47}$$

and

$$s_2(t) = -\sqrt{\frac{2E_b}{T_b}}\ cos\,(2\pi f_c t) \quad 0 \leq t \leq T_b \tag{2.48}$$

E_b and T_b are the energy and period per bit, respectively, $\phi_i(t)$ consists of a single waveform, namely,

$$\varphi_1(t) = \sqrt{\frac{2}{T_b}}cos(2\pi f_c t) \quad 0 \leq t \leq T_b \tag{2.49}$$

Using this basis signal, the BPSK signal set can be represented as,

$$S_{BPSK} = \left\{\sqrt{E_b}\,\varphi_1(t), -\sqrt{E_b}\,\varphi_1(t)\right\} \tag{2.50}$$

This signal set can be shown geometrically as in Fig. 2.26. This, so-called *constellation diagram*, provides a graphical representation of the complex envelope of each possible *symbol state*.

It should be noticed that [3, 5]: (1) The number of basis signals will always be less than or equal to the number of signals in the set. (In BPSK we have one basis signal, but two signals in the set); (2) The number of basis signals required to represent the complex modulation signal set is called the *dimension* of the vector space; (3) If there are as many basis signals as there are signals in the modulation signal set, then all the signals in the set are orthogonal to one another.

The properties of a modulation scheme that can be inferred from its constellation are as follows: (1) The bandwidth occupied by the modulation signals decreases as

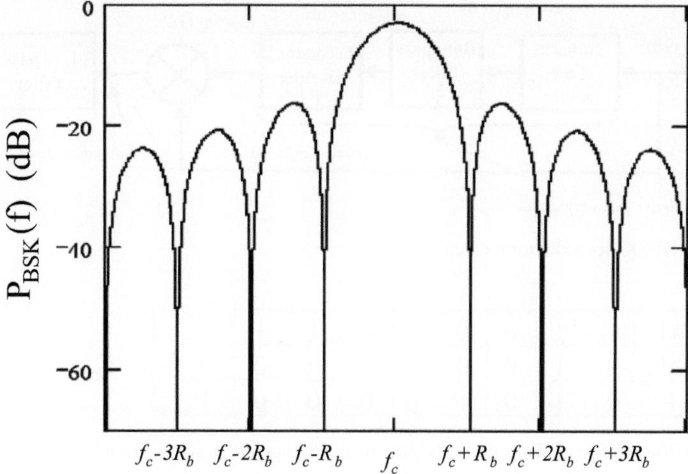

Fig. 2.27 Power spectral density for BPSK. The bandwidth, $BW = 2R_b = 2/T_b$

the number of signals/dimension increases, in particular, the more densely packed
a constellation, the more bandwidth-efficient it is; (2) The bandwidth occupied by
a modulated signal *increases* with the dimension N of the constellation; (3) The
probability of bit error is proportional to the distance between the closest points in
the constellation [3, 5].

It can be shown, that the typical power spectral density for BPSK is given by [3],

$$P_{BPSK}(f) = \frac{E_b}{2}\left[\left(\frac{sin\pi(f - f_c)T_b}{\pi(f - f_c)T_b}\right)^2 + \left(\frac{sin\pi(-f - f_c)T_b}{\pi(-f - f_c)T_b}\right)^2\right] \qquad (2.51)$$

This is plotted in Fig. 2.27. Further, the bandwidth containing the signal energy is a
function of the shape of the pulse [3]. For a rectangular pulse, 90 % of signal energy
is contained within a bandwidth $BW \sim 1.6R_b$, whereas for a raised cosine filtering
pulse with α = 0.5, 100 % of signal is contained within a $BW = 1.5R_b$ [3].

2.7.2.1   Binary Phase Shift Keying Detection

The detection of a BPSK employs a *coherent* or *synchronous* demodulation approach to extract the modulated signal, Fig. 2.28. Accordingly, information about
the carrier phase and frequency must be available at the receiver [5].

In particular, the received signal is first squared to generate a DC signal and a
varying sinusoid at twice carrier frequency. The DC is filtered out and a frequency
divider recreates the carrier which, multiplied by itself and lowpass-filtered results
in the modulation signal $m(t)$ [3].

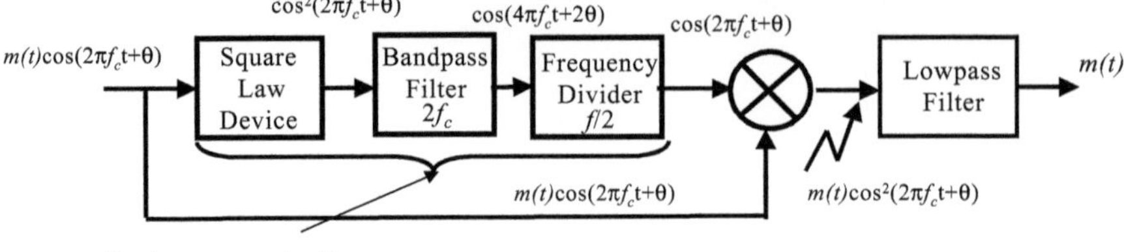

Fig. 2.28 BPSK demodulation circuit

$\{m_k\}$		1	0	0	1	0	1	1	0
$(d_k\text{-}1\}$	1	1	0	0	1	1	0	0	0
$(d_k\}$	1	1	0	1	1	0	0	0	1

$$d_k = \overline{m_k \oplus d_{k-1}}$$

Fig. 2.29 Illustration of differential BPSK. d_k is unchanged from previous symbol if $m_k = 1$, and d_k is toggled if $m_k = 0$

Assuming an additive white Gaussian noise channel (AWGN) with noise spectral density N_0, the average probability of bit error for BPSK is a function of the energy per bit, E_b is given by [3],

$$P_{e,BPSK} = Q\left(\sqrt{\frac{2E_b}{N_0}}\right) \tag{2.52}$$

where,

$$Q(x) = \int_x^\infty \frac{1}{\sqrt{2\pi}} exp(-x^2/2)dx \tag{2.53}$$

2.7.3   Differential Binary Phase Shift Keying—DBPSK

In the DBPSK modulation approach, the BPSK technique is employed in a *non-coherent* fashion to circumvent the need for a coherent reference signal at the receiver. In this scheme, the input binary sequence is first differentially encoded and then modulated using a BPSK modulator. In particular, the input binary sequence $\{m_k\}$ is used to generate a differentially encoded sequence $\{d_k\}$ by complementing the modulo-2 sum of m_k and $d_k - 1$. Figure 2.29 illustrates the differential encoding process [3].

2.7.3.1   DPSK Modulator

To implement DBPSK modulation, a one-bit delay element and logic circuit generate a differentially encoded sequence, and then the differentially-encoded bit stream

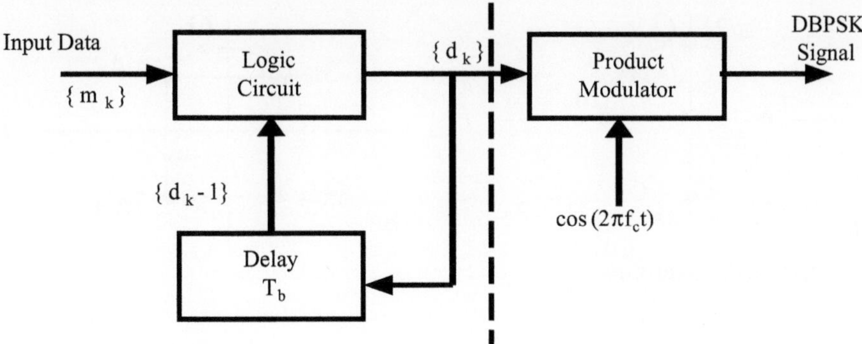

Fig. 2.30 DBPSK modulator

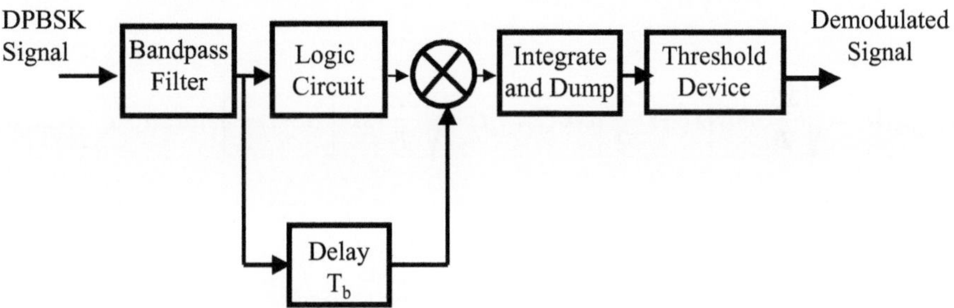

Fig. 2.31 DBPSK demodulator

is passed through a product modulator. This is depicted in the block diagram of Fig. 2.30 [3].

2.7.3.2   DPSK Demodulator

Implementation of the DPSK demodulator involves recovering the received sequence recovered through a process complementary to the modulation process, see Fig. 2.31 [3]. The approach exhibits a reduced receiver complexity, but at the expense of a 3dB inferior energy efficiency. The probability of bit errors is given by [3], [5],

$$P_{e,DPBSK} = \frac{1}{2} exp\left(-\frac{E_b}{N_0} \right) \tag{2.54}$$

2.7.4   Quadrature Phase Shift Keying Modulation—QPSK

In this modulation scheme, the phase of the carrier takes on four equally-spaced values, namely, 0, $\pi/2$, π, and $3\pi/2$, where each value of phase corresponds to a unique pair of baseband (message) bits given by [3, 5],

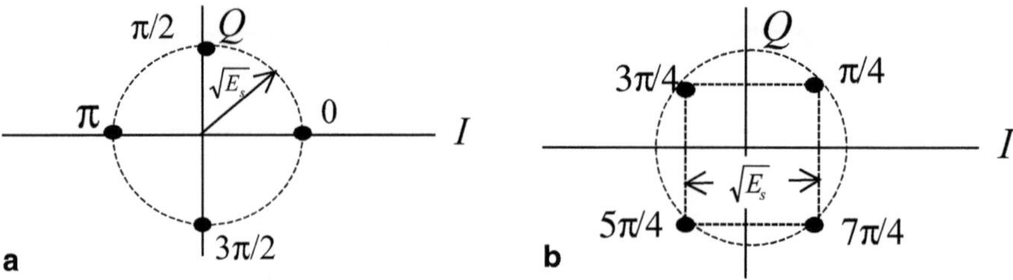

Fig. 2.32 QPSK Constellations

$$s_{QPSK}(t) = \sqrt{\frac{2E_b}{T_b}}cos(2\pi f_c t + (i-1)\frac{\pi}{2}) \quad 0 \le t \le T_s \quad i=1,2,3,4 \quad (2.55)$$

$$s_{QPSK}(t) = \sqrt{\frac{2E_b}{T_b}}cos\left[(i-1)\frac{\pi}{2}\right]cos(2\pi f_c t) - \sqrt{\frac{2E_b}{T_b}}sin\left[(i-1)\frac{\pi}{2}\right]sin(2\pi f_c t) \quad (2.56)$$

With the basis functions:

$$\varphi_1(t) = \sqrt{\frac{2}{T_s}}\,cos(2\pi f_c t) \qquad (2.57)$$

and

$$\varphi_2(t) = \sqrt{\frac{2}{T_s}}sin(2\pi f_c t) \qquad (2.58)$$

the QPSK set is,

$$S_{QPSK} = \left\{ \sqrt{E_s}\,cos\left[(i-1)\frac{\pi}{2}\right]\varphi_1(t), -\sqrt{E_s}\,sin\left[(i-1)\frac{\pi}{2}\right]\varphi_2(t)\right\} \quad i=1,2,3,4 \quad (2.59)$$

and its geometrical constellation is as shown in Fig. 2.32. The constellations are equivalent; they differ by a 45-degree rotation.

It can be shown that the power spectral density for QPSK, see Fig. 2.33, is given by [3],

$$P_{QPSK}(f) = E_b\left[\left(\frac{sin\,2\pi(f-f_c)T_b}{\pi(f-f_c)T_b}\right)^2 + \left(\frac{sin\,2\pi(-f-f_c)T_b}{\pi(-f-f_c)T_b}\right)^2\right] \qquad (2.60)$$

The null-to-null BW$=R_b$, which is half of that of a BPSK signal [3, 5].

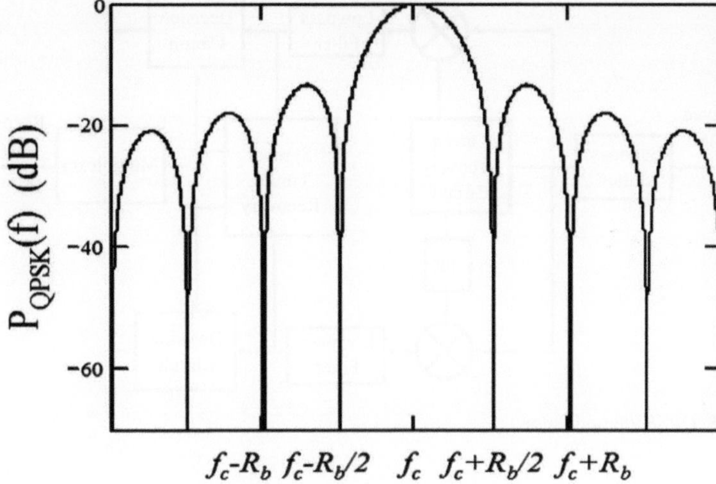

Fig. 2.33 Power spectral density for QBPSK. The bandwidth, $BW = R_b$

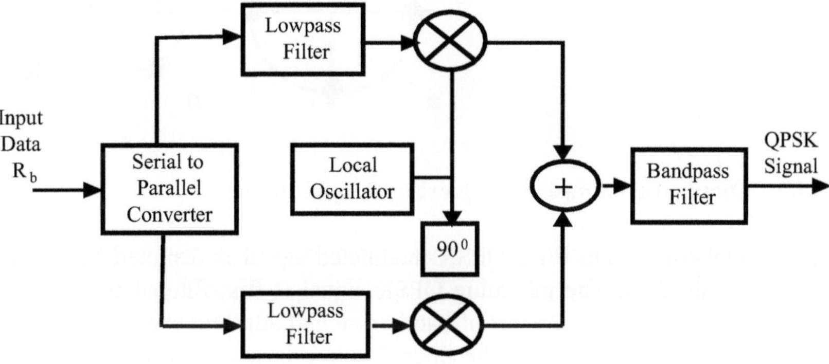

Fig. 2.34 QPSK modulator

2.7.4.1   Quadrature Phase Shift Keying Modulator

In the QPSK modulator, a unipolar bit stream $m(t)$ carrying the information signal at the rate R_b is first converted into a bipolar non-return-to-zero (NRZ) sequence. This sequence is then split into in-phase and quadrature components, $mI(t)$ and $mQ(t)$, each having a bit rate $R_b/2$. These two binary sequences, in turn, are separately modulated by two carriers, so that each is considered BPSK, and summed to produce the QPSK signal. In order to confine the power spectrum of the QPSK signal to its designated band, it is subsequently passed through a bandpass filter (BPF). A block diagram implementing the modulator is shown in Fig. 2.34 [3, 5].

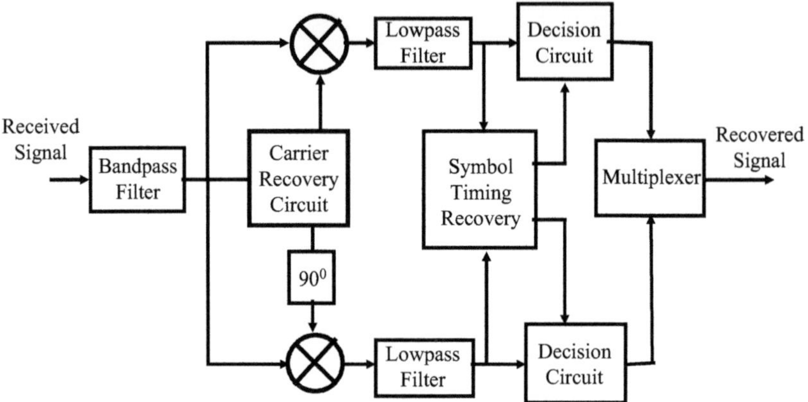

Fig. 2.35 QPSK demodulator

Fig. 2.36 $\pi/4$ QPSK constellations: **a** Possible states
when $\theta_{k-1}=n\pi/2$. **b** Possible
states when $\theta_{k-1} = n\pi/4$

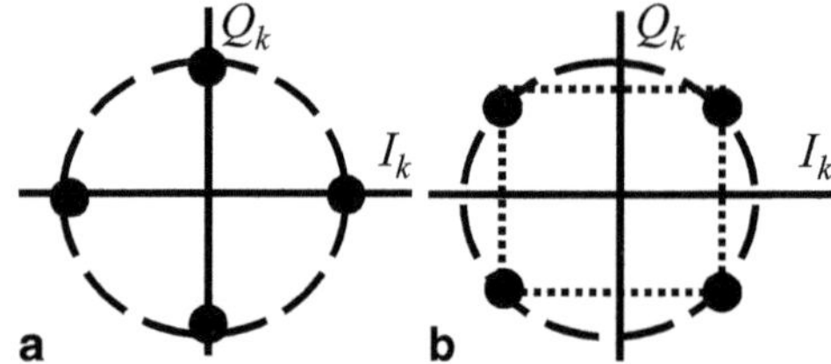

2.7.4.2   Quadrature Phase Shift Keying Demodulator

The demodulator system for a QPSK-modulated signal is depicted by the block
diagram in Fig. 2.35. The incoming QPSK signal is first filtered by a front-end
bandpass filter to remove the out-of-band noise and adjacent channel interference
accompanying it. Subsequently, the filtered output is split into two parts, each of
which is then coherently demodulated using I and Q carriers. The coherent carriers
used are recovered from the received signal using carrier recovery circuits as before
(see BPSK). Next, the outputs of the demodulators are passed through decision
circuits which generate I and Q streams. Finally, the two stream components are
multiplexed to reproduce the original binary sequence [3, 5].

2.7.5   *π/4 QPSK—Shifted Quadrature Phase-Shift Keying Modulation*

In this scheme, two QPSK constellations, shifted by $\pi/4$ with respect to each other,
are sampled to constitute the modulated signal. Selecting every successive bit by
alternating between the two constellations, ensures that there is at least a shift which
is an integer of $\pi/4$ radians between successive symbols. This guarantees that there

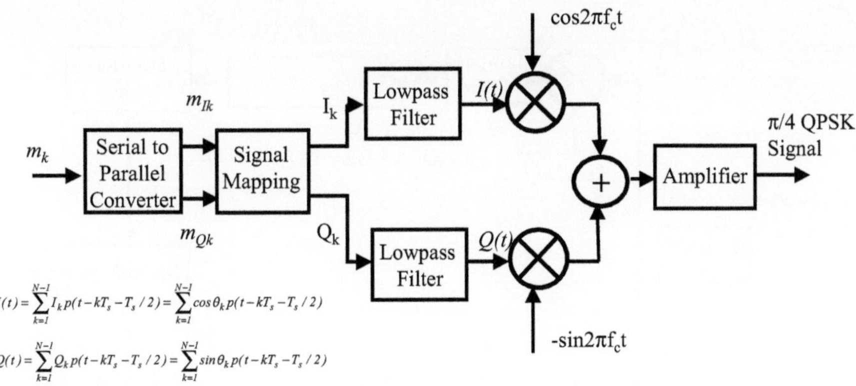

$$I(t)=\sum_{k=1}^{N-1}I_k p(t-kT_s-T_s/2)=\sum_{k=1}^{N-1}\cos\theta_k p(t-kT_s-T_s/2)$$

$$Q(t)=\sum_{k=1}^{N-1}Q_k p(t-kT_s-T_s/2)=\sum_{k=1}^{N-1}\sin\theta_k p(t-kT_s-T_s/2)$$

Fig. 2.37 π/4 QPSK modulator. Definition of variables: $I_k=\cos\theta_k=I_{k-1}\cos\phi_k-Q_{k-1}\sin\phi_k$; $Q_k\sin\theta_k=I_{k-1}\sin\phi_k+Q_{k-1}\cos\phi_k$; Phases of *kth* and *k–1st* symbols $\theta_k=\theta_{k-1}+\phi_k$

is a phase transition for every symbol, which enables a receiver to perform timing recovery and synchronization [3, 5]. Figure 2.36 depicts the two constellations.

2.7.5.1   π/4 QPSK—Shifted Quadrature Phase-Shift Keying Modulator

In the circuit implementation of this modulation scheme, Fig. 2.37, the input bit stream is first partitioned, by a serial-to-parallel converter, into $m_{I,\,k}$ and $m_{Q,\,k}$ parallel bit streams, where each symbol has a rate equal to half that of the incoming bit rate. Subsequently, each of the streams is applied to a signal mapping circuit which generates the k-th in-phase and quadrature pulses, I_k and Q_k, over the $kT<t<(k+1)T$ time interval. This is followed by the separate modulation of I_k and Q_k by two carriers which are then added to produce the π/4 QPSK signal [3]. The block diagram implementing this is shown in Fig. 2.37.

2.7.5.2   π/4 QPSK—Shifted Quadrature Phase-Shift Keying Demodulator

Demodulation of a π/4 QPSK signal proceeds as follows (the numbers correspond to those in diagram of Fig. 2.38): (1) The incoming π/4 QPSK signal is first quadrature-demodulated using two LO signals with the frequency identical to the transmitted carrier; (2) The output product is applied to lowpass filters, whose output is sampled at the maximum amplitude every T_s seconds; (3) The two sequences are then passed through differential decoders; (4) The output of differential decoder may be expressed as $x_k=\cos(\phi_k-\phi_{k-1})$, $y_k=\sin(\phi_k-\phi_{k-1})$; (5) The output of the differential decoder is applied to the decision circuit, which uses the table in Fig. 2.36(b) to determine the output bit as follows: $S_I=1$, if $x_k>0$ or $S_I=0$, if $x_k<0$ $S_Q=1$, if $y_k>0$ or $S_Q=0$ if $y_k<0$.

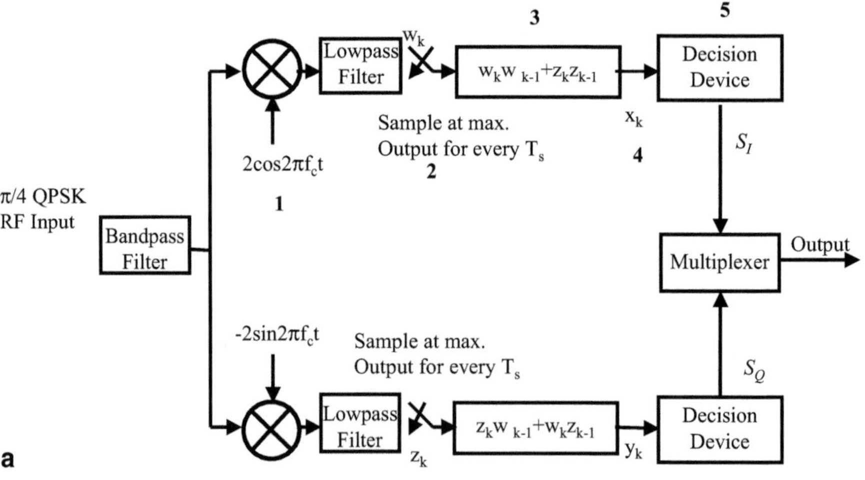

Information bits mI_k, mQ_k	Phase Shift ϕ_k
11	$\pi/4$
01	$3\pi/4$
00	$-3\pi/4$
10	$-\pi/4$

b

Fig. 2.38 a $\pi/4$ QPSK demodulator. Definition of variables: In-Phase: $w_k = \cos(\phi_k - \gamma)$; In Quadrature: $z_k = \sin(\phi_k - \gamma); \phi_k = \tan^{-1}(Q_k / I_k); \gamma$ due to noise. **b** Carrier phase shift corresponding to various input bit pairs

2.7.5.3 $\pi/4$ Shifted QPSK—Demodulator with IF Differential Detector

This scheme circumvents the need for local oscillators (LO), see Fig. 2.38, by using a delay line and 90° phase shifter to implement the quadrature, and two phase detectors, see Fig. 2.39. The operation is as follows (the numbers correspond to those in diagram of Fig. 2.39): (1) The received signal is converted to intermediate frequency (IF) and band-pass filtered; the bandpass filter is designed to match the transmitted pulse shape to preserve carrier phase and minimize noise power; (2) The received IF is differentially decoded using delay line and two mixers; (3) Similar to baseband differential detector already discussed [3].

2.7.5.4 $\pi/4$ QPSK Demodulator: FM Discriminator

This scheme operates as follows (the numbers correspond to those in diagram of Fig. 2.40): (1) The input signal is first filtered using a bandpass filter (BPF) that

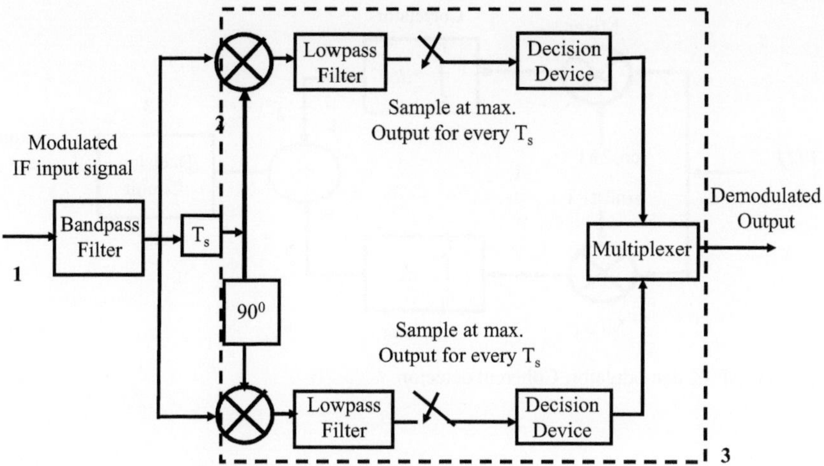

Fig. 2.39 $\pi/4$ QPSK Demodulator IF Differential Detector

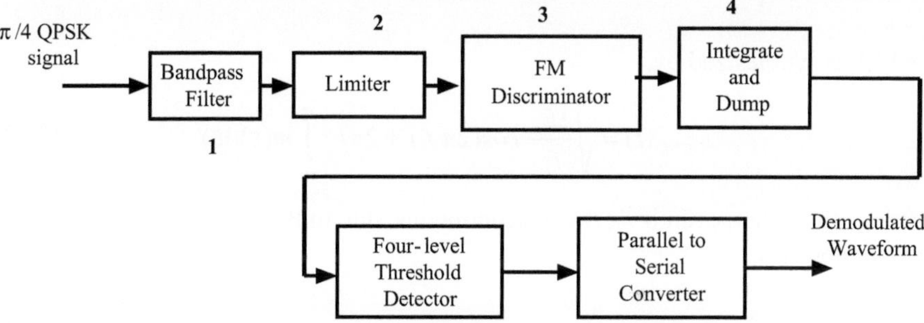

Fig. 2.40 $\pi/4$ QPSK Demodulator: FM Discriminator

is matched to the transmitted signal; (2) The filtered signal is passed by a hard-limiter to minimize fluctuations in its envelope; (3) The FM discriminator extracts the instantaneous frequency; (4) The extracted signal is integrated over each period giving the phase difference between two sampling instants [3].

2.7.6   *Binary Frequency-Shift Keying—BFSK*

In this scheme, the frequency of a constant-amplitude carrier is switched between two values (bits 1 and 0). Then, depending on how frequency changes, there may be discontinuous phase or continuous phase between bits. The BFSK signal is given by [3],

$$s_{BFSK}(t) = \sqrt{\frac{2E_b}{T_b}} cos(2\pi f_c + 2\pi\Delta f_c)t \quad 0 \leq t \leq T_b \quad (binary\ 1) \quad (2.61)$$

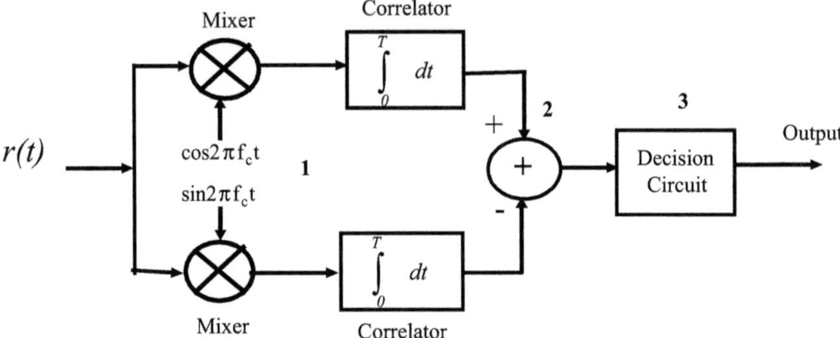

Fig. 2.41 BFSK demodulator: Coherent detector

or

$$s_{BFSK}(t) = \sqrt{\frac{2E_b}{T_b}}\, cos(2\pi f_c - 2\pi\Delta f)t \quad 0 \le t \le T_b \quad (binary\ \ 0) \qquad (2.62)$$

which is synthesized by,

$$s_{BFSK}(t) = \sqrt{\frac{2E_b}{T_b}}\, cos(2\pi f_c t + 2\pi k_f \int_{-\infty}^{\infty} m(\xi)d\xi) \qquad (2.63)$$

It is noticed that, even if $m(t)$ is discontinuous, due to the integration, the phase is continuous [3].

2.7.6.1   BFSK Modulator

The most common way to effect BPSK modulation is direct FM, that is, modulating the frequency of an oscillator with the message signal [3].

2.7.6.2   BFSK Demodulator: Coherent Detector

This demodulation scheme is implemented as shown in Fig. 2.41, and operates as follows (the numbers correspond to those in diagram): (1) Two correlators are applied to reference signals that are generated locally and that are *coherent*; (2) The difference of the correlator outputs is then compared in a threshold comparator; (3) The output signal is the result of the comparison, in particular, if the difference signal has a value greater than the threshold, it is classified as "1", otherwise it is classified as a "0." The probability of bit error is given by [3],

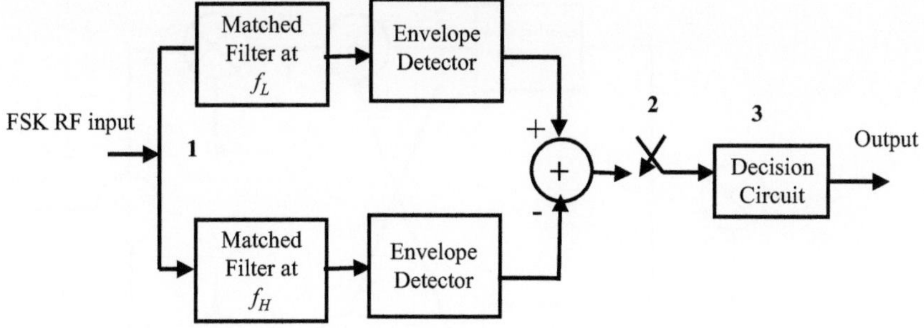

Fig. 2.42 BFSK demodulator: Noncoherent detector

$$P_{e,BPSK} = Q\left(\sqrt{\frac{E_b}{N_0}}\right) \tag{2.64}$$

2.7.6.3   BFSK Demodulator: Noncoherent Detector

To demodulate the BFSK signal, the block diagram shown in Fig. 2.42 is employed; it operates as follows (the numbers correspond to those in diagram): (1) The incoming signal is first split into two paths, namely, a lower path where it is passed through a filter that is matched to the frequency f_H, and an upper path where it is passed by a filter matched to the frequency f_L; (2) The matched filters function as bandpass filters centered at f_H and f_L, and are passed through envelop detectors; (3) The difference of the outputs of the envelope detectors is then sampled at times $t = kT_b$; (4) Depending on the magnitude of the sampled value, the decision circuit decides the bit classification as 1 or 0. The probability of bit error rate is [3],

$$P_{e,FSK,NC} = \frac{1}{2} exp\left(\frac{-E_b}{2N_0}\right) \tag{2.65}$$

2.7.7   *Minimum Shift Keying—MSK*

The MSK modulation approach entails producing a peak deviation in the frequency, equal to one-quarter the bit rate; since it is a special type of continuous phase-frequency shift keying (CPFSK), MSK is also denoted continuous FSK, with modulation index of 0.5. In general, the modulation index for this scheme is defined as: $k_{FSK} = (2\Delta F)/R_b$, where ΔF is the peak deviation; R_b is bit rate. Mobile radio systems frequently use MSK due to its spectrally efficient modulation scheme. MSK possesses: (1) Constant envelope; (2) Good BER performance; (3) Self-synchronizing capability [3].

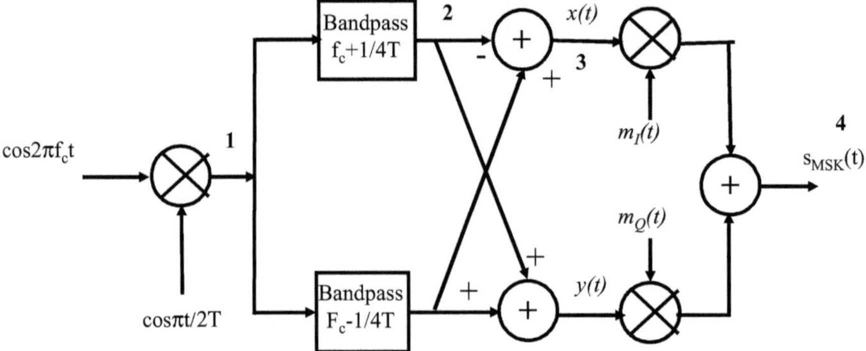

Fig. 2.43 MSK modulator

The MSK signal is given by,

$$s_{MSK}(t) = \sum_{i=0}^{N-1} m_{I_i}(t)\,p(t-2iT_b)\cos 2\pi f_c t + \sum_{i=0}^{N-1} m_{Q_i}(t)\,p(t-2iT_b-T_b)\sin 2\pi f_c t \qquad (2.66)$$

where,

$$p(t) = \begin{cases} \cos\left(\dfrac{\pi t}{2T_b}\right) & 0 \le t \le 2T_b \\[2ex] 0 & elsewhere \end{cases} \qquad (2.67)$$

where m_I and m_Q are the even and odd bits of a bipolar serial data stream with values $+/-1$; they are fed at the rate $R_b/2$ [3]. The MSK waveform may be represented by,

$$s_{MSK}(t) = \sqrt{\frac{2E_b}{T_b}}\,\cos(2\pi f_c t - m_{I_i}(t)m_{Q_i}(t)\frac{\pi t}{2T_b} + \varphi_k) \qquad (2.68)$$

MSK may be visualized as a special type of continuous phase FSK, in which ϕ_κ is 0 or π depending on whether $m_I(t)$ is 1 or -1 [3].

2.7.7.1   MSK Modulator

The block diagram of the MSK modulator is shown in Fig. 2.43, and its operation is as follows (the numbers correspond to those in diagram): (1) The carrier is multiplied with $\cos(\pi t/2T)$ to produce two-phase-coherent signals at $f_c+1/4T$ and $f_c-1/4T$; (2) The two FSK signals are separated using two bandpass filters centered at the respective frequencies; (3) And combined to form in-phase and quadrature carrier components, $x(t)$ and $y(t)$, respectively; (4) These carriers are multiplied with even and odd bit streams, $m_I(t)$ and $m_Q(t)$ to produce $s_{MSK}(t)$.

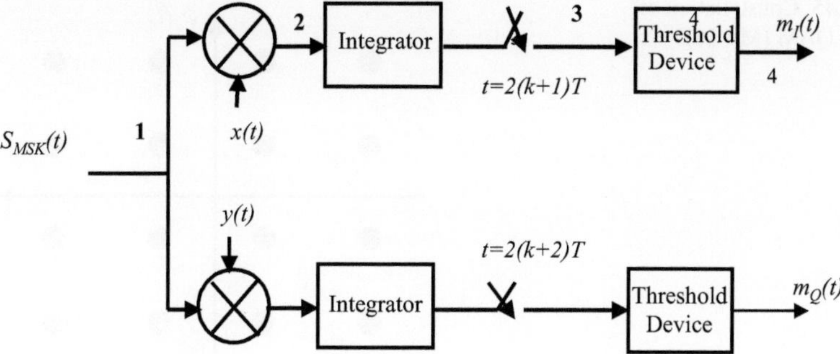

Fig. 2.44 MSK demodulator

2.7.7.2   MSK Demodulator

The block diagram of the MSK demodulator is shown in Fig. 2.44, and its operation is as follows (the numbers correspond to those in diagram): (1) First, the received incoming signal is multiplied by the respective in-phase and quadrature carriers $x(t)$ and $y(t)$; (2) The output of the multipliers are integrated over two bit periods and dumped to a decision circuit at the end of the period; (3) Based on the level of the signal, a threshold detector decides its classification as 1 or 0; (4) The output data streams correspond to $m_I(t)$ and $m_Q(t)$, which are offset to obtain the demodulated signal.

2.7.8   *M-ary QAM—Quadrature Amplitude Modulation*

In this scheme, the carrier amplitude is allowed to vary together with the phase. The signal is given by,

$$s_i(t) = \sqrt{\frac{2E_{min}}{T_s}}a_i cos(2\pi f_c t) + \sqrt{\frac{2E_{min}}{T_s}}b_i sin(2\pi f_c t)$$

$$0 \le t \le T \quad i = 1,2,....,M \tag{2.69}$$

where E_{min} is the energy of the signal with the lowest amplitude, and a_i and b_i are a pair of independent integers according to location of a signal point. The pertinent constellation for, e.g., M = 16, is as given in Fig. 2.45.

2.7.9   *OFDM—Orthogonal Frequency Multiplexing*

The OFDM approach to signal transmission provides power-efficient signaling for a large number of users in the same channel. In this scheme, multiple sub-carriers

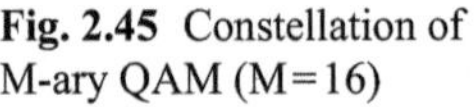

Fig. 2.45 Constellation of M-ary QAM (M = 16)

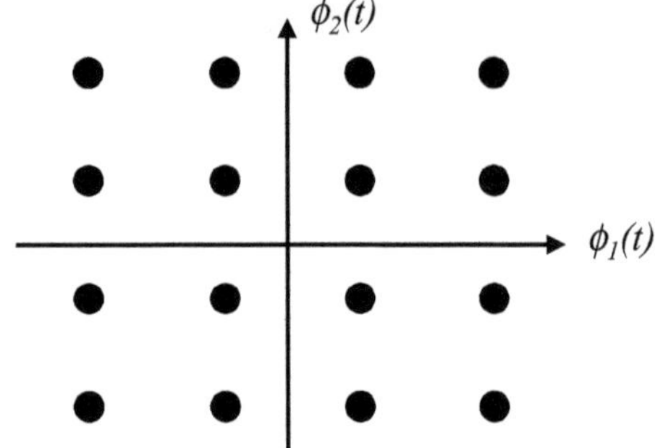

Fig. 2.46 a OFDM modulator block diagram. **b** Carriers modulated by rectangularly-shaped data pulses are densely spaced and orthogonal in frequency. *After* [1]

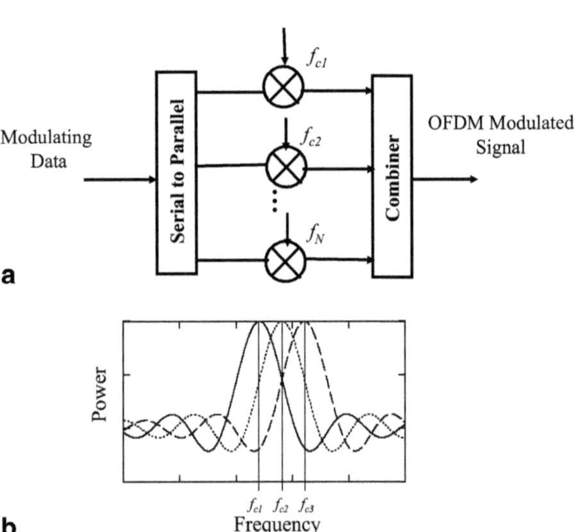

are employed to transmit signal bits in parallel for very high throughput. The spectra of different sub-channels can partly overlap and pulse shaping is not necessary. OFDM is usually combined with QAM or M-PSK [1] (Fig. 2.46).

2.7.10   *Direct Sequence Spread Spectrum Modulation*

The idea behind the direct sequence (DS) spread spectrum (SS) scheme, depicted in Fig. 2.47, is to greatly expand or spread the carrier spectrum relative to the information rate [6]. The spread spectrum signal attains an anti-jamming capability by forcing the jammer to deploy the transmitted power over a much wider bandwidth than would be necessary for a conventional system [6]. In other words, for a given jammer power, the jamming power spectral density is reduced in proportion to the ratio of the spread bandwidth B_s to the un-spread bandwidth B.

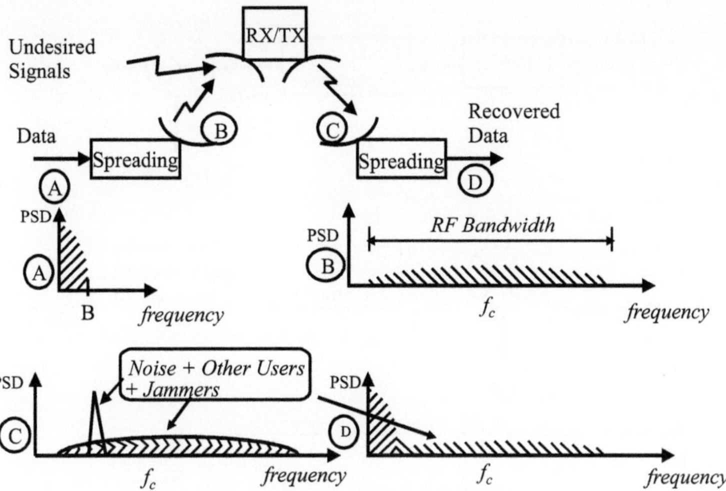

Fig. 2.47 Direct sequence spread spectrum modulation concept

2.7.10.1   Direct Sequence Spread Spectrum Modulation/Demodulation

DS/SS modulator and demodulator block diagrams are shown in Fig. 2.48. To produce an SS signal, the input data is passed through a data encoder and a data modulator, which results in the signal message $m(t)$. $m(t)$ is subsequently applied to a mixer, where it is multiplied by the product of the carrier and a pseudorandom sequence (a random-like signal of ± 1 $p(t)$) to produce [6],

$$V_{SS}(t) = \sqrt{\frac{2E_s}{T_s}} m(t) p(t) \cos(2\pi f_c t + \theta) \tag{2.70}$$

where,

$$m(t) = \sum_{m=-\infty}^{\infty} a_m p_m(t - mT_b) \tag{2.71}$$

and

$$p(t) = \sum_{n=-\infty}^{\infty} c_n p'(t - nT_c) \tag{2.72}$$

The DS/SS system is employed in Code Division Multiple Access (CDMA), where each user has pseudo-noise (PN) sequence, and multiple users share the same BW; the signals of one user appearing as noise to others. A feedback shift register with

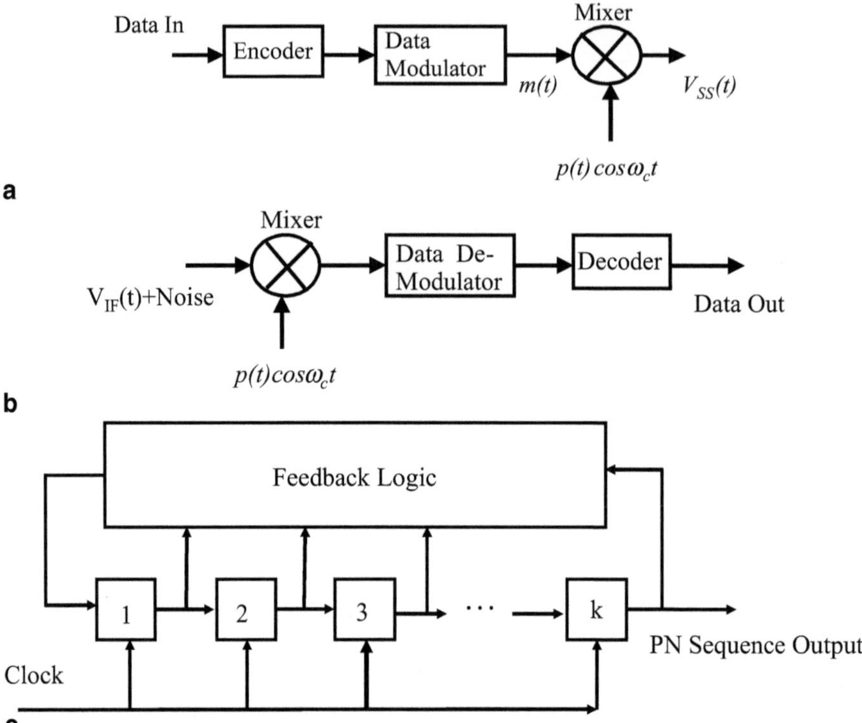

Fig. 2.48 Direct sequence spread spectrum: **a** Modulator. **b** Demodulator. **c** PN sequence generator diagram

k stages is employed for generating the PN sequence, (2.72), Fig. 2.48c. For a shift register of k stages, the PN sequence length is $N = 2^k - 1$ [6]. The measure of interference rejection of a DS/SS system is given by the so-called *processing gain*, defined by,

$$PG_{DS} = \frac{T_b}{T_c} = \frac{R_c}{R_b} \tag{2.73}$$

2.7.11 *Frequency Hopping Spread Spectrum Modulation/Demodulation*

The frequency hopping (FH) spread spectrum scheme, Fig. 2.49, involves the periodic change of the transmission frequency [6]. In this scheme a sequence of modulated data bursts with time-varying, pseudo-random carrier frequencies, is transmitted. In FH the processing gain is given by,

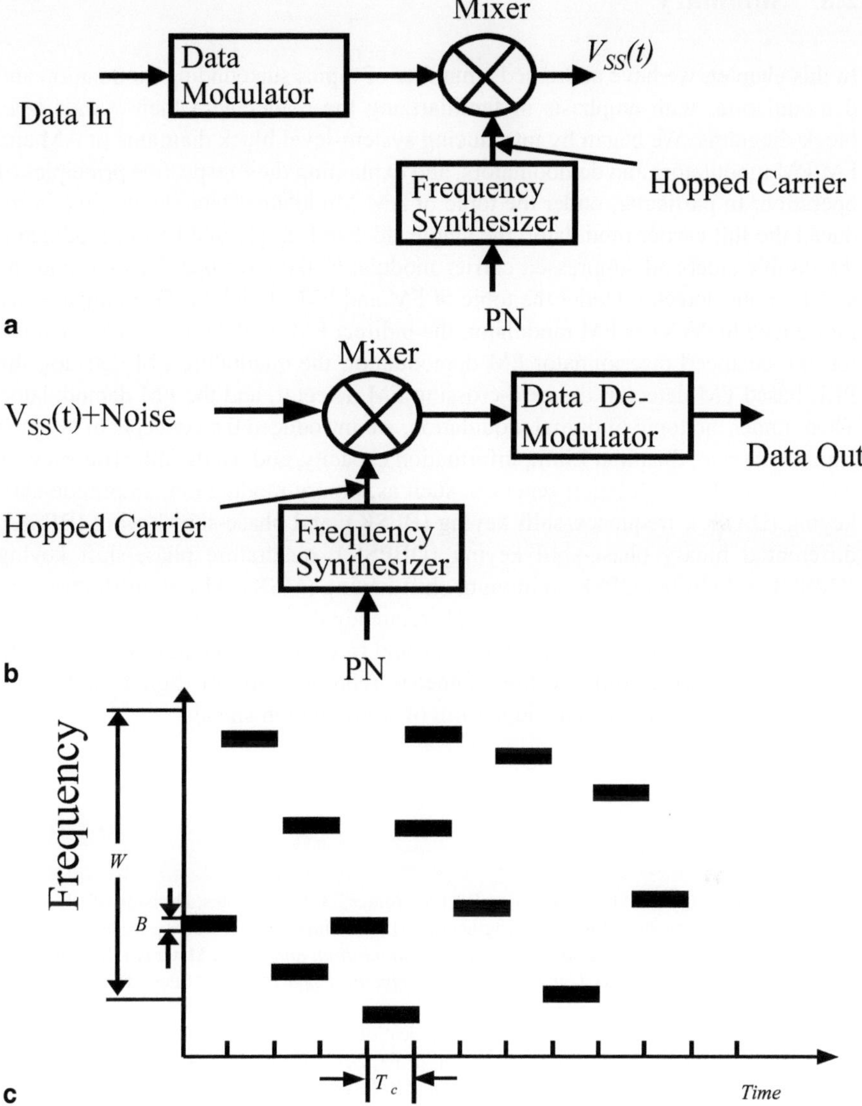

Fig. 2.49 Frequency hopping spread spectrum: **a** Modulator. **b** Demodulator. **c** Frequency hopping versus time. T_c is the duration of time in a given frequency band

$$PG_{FHS} = \frac{W}{B} \tag{2.74}$$

where, W is the frequency range through hopping can occur, and B is the bandwidth of the band being hopped [6].

2.8   Summary

In this chapter, we have addressed a number of topics surrounding modulation and demodulation, with emphasis in familiarizing the reader with their *system-level* block diagrams. We began by introducing system-level block diagrams of AM and FM/PM modulators and demodulators, and explaining their respective principles of operation. In particular, under the topic of AM Modulator/Demodulator, we introduced the full carrier modulator, the single sideband suppressed carrier modulator, the double sideband suppressed carrier modulator, the envelope detector, and the synchronous detector. Under the topic of FM and PM Modulator/Demodulator, we introduced theVCO as FM modulator, the indirect FM modulator, the PM modulator, the balanced discriminator FM demodulator, the quadrature FM detector, the PLL-based FM detector, the zero-crossing FM detector, and the PM demodulator. Then, under the topic of digital modulation, we introduced the concepts of Nyquist Limit, data rate, Shannon Limit, information capacity, and bandwidth efficiency, as well as specific modulation schemes, such as, binary modulation, amplitude-shift keying (BASK), frequency-shift keying (BFSK), and phase-shift keying (BPSK), differential binary phase-shift keying (DBPSK), quadrature phase-shift keying (QPSK), $\pi/4$ shifted QPSK, minimum shift keying (MSK), M-ary quadrature amplitude modulation (QAM), orthogonal frequency division multiplexing (OFDM), direct sequence spread spectrum (DS/SS), and frequency hopping spread spectrum (FH/SS). We also introduced the geometric representation of digital modulation schemes and the complex envelope form of a modulation signal.

References

1. J. Dąbrowski, Course "Introduction to RF Electronics," Lecture Notes; Division of Electronic Devices, Department of Electrical Engineering (ISY), Linköping University, 2006.
2. Mischa Schwartz, *Information Transmission, Modulation, and Noise*, McGraw-Hill, 1970.
3. T. S. Rappaport, *Wireless Communications: Principles and Practice*, Second Ed., Prentice-Hall, Inc. 2002.
4. H. J. De Los Santos, Lecture Notes, "Communications Circuits," Course EE115D, Department of Electrical Engineering, University of California, Los Angeles (UCLA), Fall Quarter, 1990.
5. Q. Gu, *RF System Design of Transceivers for Wireless Communications*, Springer, 2005.
6. R. C. Dixon, *Spread Spectrum Systems with Commercial Applications*, Third Edition, John Wiley & Sons, Inc., New York, 1994.
7. Available: [Online]: http://www.clee.freehomepage.com/teaching.html/

Chapter 3
Typical System Performance Parameters

Abstract In this chapter, we address a number of topics surrounding system performance parameters. We begin by discussing transmitter's maximum output power, modulation accuracy, and adjacent and alternate channel power. Then, on the receiver side, we discuss the topics of sensitivity, noise figure, selectivity, image rejection (IR), dynamic range, and spurious-free dynamic range. Next, we introduce a variety of system building blocks, in particular, their description and typical models for systems analysis. These include, transmission lines, amplifiers, mixers, filters, oscillators, and frequency multipliers.

3.1 Transmitter

In general, typical transmitter requirements include: (1) Maximum output power; (2) Modulation accuracy; and (3) Adjacent and Alternate Channel Power. The latter two are addressed next.

3.1.1 Modulation Accuracy

Modulation accuracy is the fidelity of the transmitted waveform, as measured by the *Error Vector Magnitude* EVM or *Waveform Quality Factor* ρ [1]

EVM is the difference between the actual symbol location and the theoretical symbol location on the modulation vector constellation diagram, Fig. 3.1 [1].

EVM is quantified as follows [1]:

$$EVM = \left\{ \frac{E\left\{|\bar{e}|^2\right\}}{E\left\{|\bar{a}|^2\right\}} \right\}^{1/2} \tag{3.1}$$

E{.} is the expectation value of ensemble averages. Waveform Quality Factor, ρ, on the other hand, is the correlation coefficient between the actual waveform $Z(t)$ and the ideal waveform $R(t)$, and is quantified as,

H. J. De Los Santos et al., *Radio Systems Engineering,*
DOI 10.1007/978-3-319-07326-2_3, © The Authors 2015

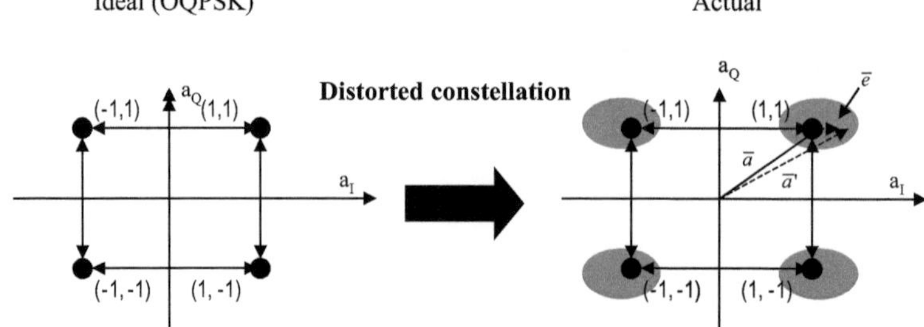

Fig. 3.1 Constellation distortion due to noise. Example of how ideal *OQPSK* constellation may be distorted en-route from transmitter to receiver

$$\rho = \frac{\left| \sum_{k=1}^{M} R_k Z_k^* \right|}{\sum_{k=1}^{M} |R_k|^2 \cdot \sum_{k=1}^{M} |Z_k|^2} \tag{3.2}$$

For negligible cross-correlation between the ideal and the error signal, it can be shown that [1],

$$\rho \cong \frac{1}{1 + EVM^2} \tag{3.3}$$

In general, *EVM* has many contributions, namely, Inter-symbol Interference, Close-in Phase Noise, Carrier Leakage (Carrier Feed-Through-CFT), I & Q Imbalance, PA Nonlinearity, In-Channel Bandwidth Noise, etc. The total EVM encompassing these effects is given by [1],

$$EVM = \sqrt{\sum_k EVM_k^2} = \sqrt{EVM_{ISI}^2 + EVM_{CFT}^2 + \sum_i EVM_{Nphasse,i}^2 + ...} \tag{3.4}$$

3.1.2 Adjacent and Alternate Channel Power

This is the ratio of the power integrated over an assigned bandwidth in the adjacent/alternate channel, to the total transmission power, Fig. 3.2. It is quantified by [1],

$$ACPR = \frac{\int_{f_1}^{f_1 + \Delta B_{ACP}} SPD(f)df}{\int_{f_0 - BW/2}^{f_0 + BW/2} SPD(f)df} \tag{3.5}$$

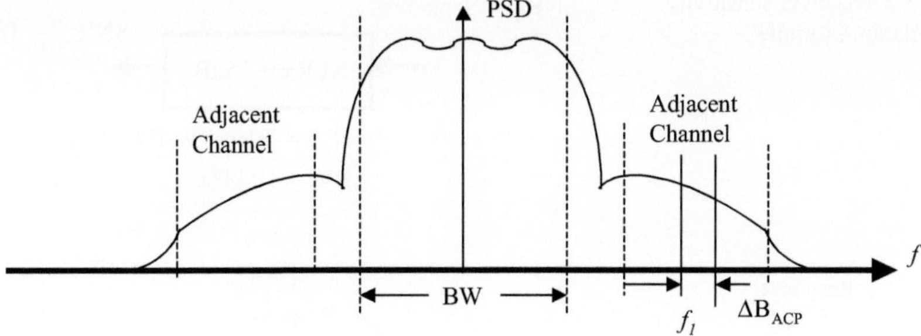

Fig. 3.2 Description of adjacent channel power

3.2 Receiver

In general, typical receiver requirements include: (1) Sensitivity; (2) Noise Figure; (3) Selectivity; (4) Image Rejection (IR); (5) Dynamic Range; and (6) Spurious Free Dynamic Range (SFDR). Next, we define these.

3.2.1 Sensitivity

The sensitivity of a receiver is the minimum input carrier voltage that will produce a specified signal-to-noise power ratio (SNR) at the output of the IF section. This parameter captures the ability of the receiver to detect a signal with given SNR, or in other words, the lowest input signal power that guarantees the needed SNR at the Rx output for a given power. For example, consider the receiver in Fig. 3.3. If $SNR_{out} = 18$ dB $= 63$, BW $= 10$ kHz, NFRx $= 15$dB $= 32$, then, $SNR_{in} = Sens/N_{ref}$, $N_{ref} = 4 \times 10^{-21} \times 10 \times 10^3 = 4 \times 10^{-17}$ W, and Sens $= SNR_{out} \times NFRx \times N_{ref} = 2 \times 10^{-13}$ W, which is equal to 2.5 μV with a 75 Ω antenna.

3.2.2 Noise Figure

The noise Figure (NF) is a measure of the degradation of the SNR between the input (i.e., antenna) and output (i.e., baseband).

It is given by [2],

$$NF = \frac{input\ SNR}{output\ SNR} = \frac{P_{si}/P_{ni}}{P_{so}/P_{no}} \tag{3.6}$$

where P_{si}, P_{sn}, and P_{so}, P_{no} are the power and noise at the input and the output, respectively. The value of NF is often expressed in decibels: NFdB $= 10 \log_{10}$ NF.

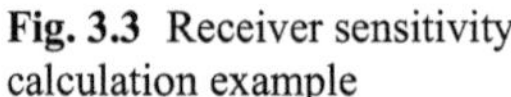

Fig. 3.3 Receiver sensitivity calculation example

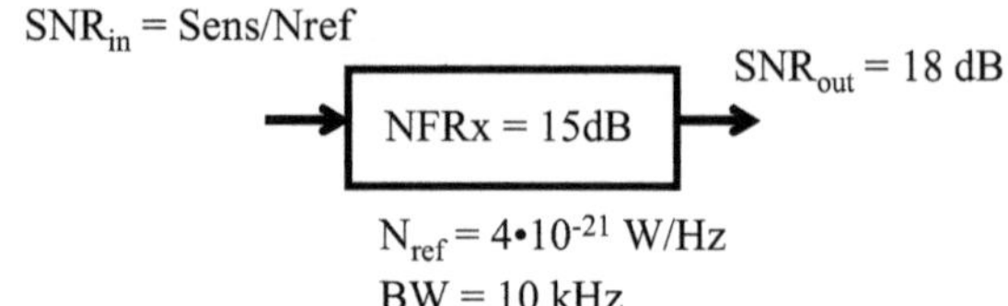

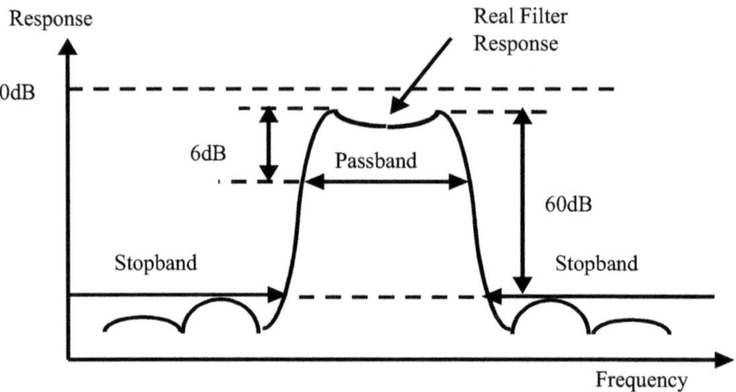

Fig. 3.4 Filter response used in definition of shape factor. For receivers the *passband* is taken to be the bandwidth between the points where the response has fallen by 6 dB, i.e. where it is 6 dB down or -6 dB

3.2.3   *Selectivity*

This is the ability to discriminate a wanted signal against interferers (images), expressed by the ratio of the bandwidths of the stop band to the pass band, see Fig. 3.4. In other words, it is the parameter that determines whether the receiver is able to pick out the wanted signal from all the other ones around it. This is determined primarily by the filters in the IF section.

The quantitative measure of selectivity is the *Shape Factor* (SF), defined with respect to Fig. 3.4 by,

$$SF = \frac{BW - 60dB}{BW - 6dB} \tag{3.7}$$

Many of the filters used in radio receivers have very high levels of performance and enable radio receivers to select individual signals even in the presence of many other close-by signals. For example, a filter with a passband of 3 kHz at -6 dB and a passband of 6 kHz at -60 dB for the stop band, would have a shape factor of 2:1.

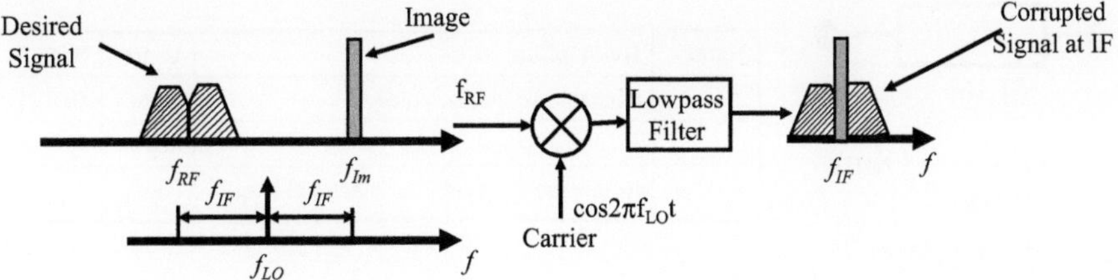

Fig. 3.5 Illustration of image problem

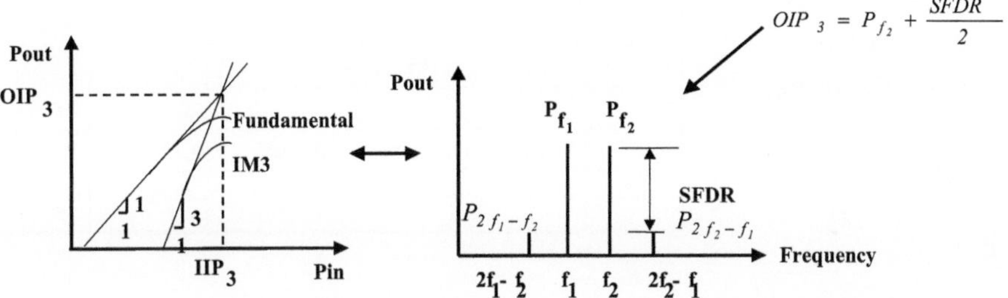

Fig. 3.6 Power transfer and power spectral relationships illustrating *SFDR* concept

3.2.4   Image Rejection (IR)

This is the ratio of the receiver (Rx) gain for the wanted signal to the Rx gain for the image, measured in dB. The image problem is illustrated in Fig. 3.5. Signals located at the same distance from the local oscillator that the desired RF signal, will also be identically down-converted with the RF, such that they will both land at the IF. If the image is stronger than the desired RF signal, it will corrupt it, making it difficult to discern the desired RF signal.

3.2.5   Dynamic Range

This is the ratio between largest input signal tolerated by the Rx and the sensitivity measured in dB.

3.2.6   Spurious Free Dynamic Range (SFDR)

This is the combined measure of linearity and noise performance, Fig. 3.6. The SFDR is the sum of the output power of, say, tone f_2, and half the difference, in dB, in the output power of a tone f_2 and the output power at the 3rd-order product.

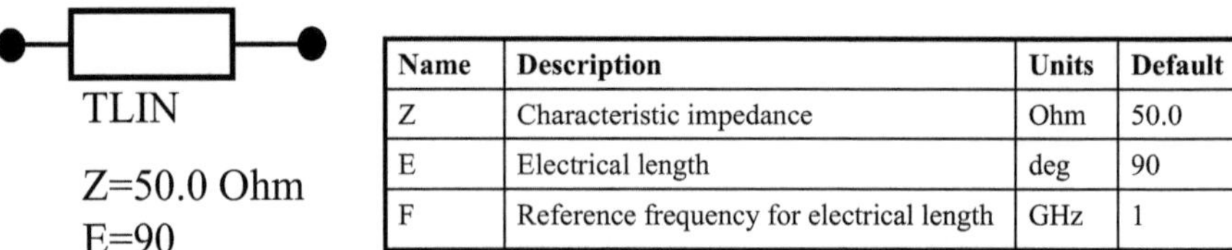

Name	Description	Units	Default
Z	Characteristic impedance	Ohm	50.0
E	Electrical length	deg	90
F	Reference frequency for electrical length	GHz	1

Name	Description	Units	Default
Z	Characteristic impedance	Ohm	50.0
L	Physical length	mil	1000.0
K	Effective dielectric constant	None	2.1
A	Attenuation	dB/meter	0.0001
F	Frequency for scaling attenuation	GHz	1
TanD	Dielectric loss tangent	None	0.0002

Fig. 3.7 Parameters describing transmission line. **a** Ideal TL **b** Physical TL

3.3   System Components

The fundamental components of transmitters and receivers may be surmised from inspection of the previous figures and discussions. They are:

1. Transmission Lines
2. Amplifiers
3. Mixers
4. Filters
5. Oscillators
6. Frequency Multipliers

Next, a description of these components, as it is pertinent for systems engineering analyses, is given.

3.3.1   Transmission Lines

Transmission lines (TLs) are employed to interconnect the system's building blocks. Ideally, they should do so without changing the amplitude or power, and the frequency content or bandwidth, of the signal they convey. We will present two models, namely, an ideal and a physical TL.

The ideal TL has no attenuation and an infinite bandwidth, Fig. 3.7.

3.3.2   Amplifiers

Amplifiers are employed to boost the power level of a signal. Ideally, an amplifier should do so without altering the signal's frequency content, regardless of the input signal amplitude. In reality, however, the input signal is almost always the sum of a desired signal and one or more interfering signals (also called interferers). As a result, since the amplifier gain is a nonlinear function of input power, these multiple signals interact with the amplifier's nonlinearities to produce additional spectral components, not contained in the original input signal. The characterization and model of an amplifier, capturing its nonlinear behavior, is addressed next.

The input-output transfer curve of an amplifier may be represented as,

$$y(t) = F(x(t)) = \sum_{n=1}^{\infty} a_n x^n(t) \cong \sum_{n=1}^{N} a_n x^n(t) \tag{3.8}$$

which, retaining the first four terms, may be written as,

$$y(t) = a_0 + a_1 x(t) + a_2 x^2(t) + a_3 x^3(t) \tag{3.9}$$

In this equation, the first two terms represent the linear part, and the third and fourth terms the nonlinear part. This nonlinearity gives rise to the following problems.

3.3.2.1   Gain Compression and Desensitization

Consider the presence of the following signals at the input of an amplifier,

$$S_d(t) = A_d \cos 2\pi f_0 t \tag{3.10}$$

$$S_{I1}(t) = A_{I1} \cos 2\pi_1 ft \tag{3.11}$$

$$S_{I2}(t) = 0 \tag{3.12}$$

where, S_d, S_{I1} and S_{I2} represent the desired and two possible interfering signals, the latter with zero amplitude. Then, substituting these into (3.9) one gets,

$$y(t) = a_1[S_d(t) + S_{I1}(t) + S_{I2}(t)] + a_2[S_d(t) + S_{I1}(t) + S_{I2}(t)^2] + a_3[S_d(t) + S_{I1}(t) + S_{I2}(t)]^3 + \dots$$

$$\tag{3.13}$$

This, upon insertion of (3.9, 3.10, 3.11), yields,

$$\begin{aligned}
y(t) = a_1 &\left[A_d \cos 2\pi f_0 t + A_{I1} \cos 2\pi f_1 t \right] \\
+ a_2 &\left[A_d^2 \cos^2 2\pi f_0 t + A_{I1}^2 \cos^2 2\pi f_1 t + 2 A_d A_{I1} \cos 2\pi f_0 t \cos 2\pi f_1 t \right] \\
+ a_3 &\left[A_d^3 \cos^3 2\pi f_0 t + A_{I1}^3 \cos^3 2\pi f_1 t + 3 A_d^2 A_{I1} \cos^2 2\pi f_0 t \cos 2\pi f_1 t \right. \\
&\left. + 3 A_d A_{I1}^2 \cos 2\pi f_0 t \cos^2 2\pi f_1 t \right] + \dots
\end{aligned} \tag{3.14}$$

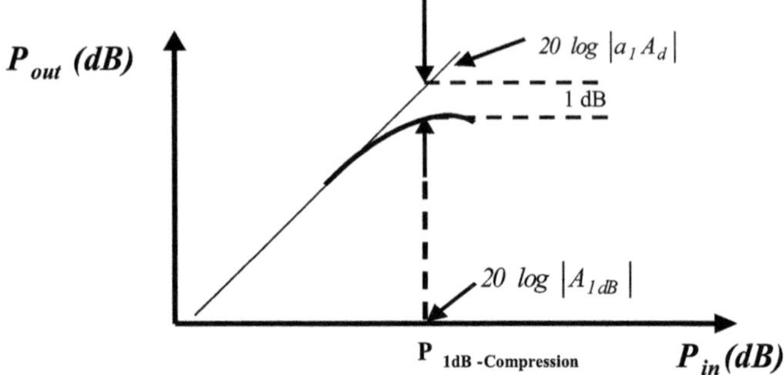

Fig. 3.8 Description of 1 dB compression point

which may be simplified as,

$$y(t) = a_1 A_d \left[1 + \frac{3a_3}{4a_1} A_d^2 + \frac{3a_3}{2a_1} A_{I1}^2 \right] \cos 2\pi f_0 t + \dots \tag{3.15}$$

Now, when the *interference is absent* and the desired signal $A_d << 1$, the *small signal gain* is equal to a_1, since for a mildly nonlinear system all the terms above will be negligibly small, if compared to the first term $a_1 A_d \cos 2\pi f_0 t$. However, as the amplitude of the desired signal increases, the gains, as determined by the term within brackets, will vary with the input signal level. In particular, the second term in the brackets in (3.15), $3a_3 A_d^2 / 4a_1$, gradually becomes significant. Then, if the sign of a_3 is opposite that of a_1, the output, $y(t)$, will be smaller than that predicted by linear theory giving rise to the phenomenon called *gain compression*. The compressive gain G_c in decibel (dB) is given by:

$$G_c = 20 \log \left| a_1 \left(1 + \frac{3a_3}{4a_1} A_d^3 \right) \right| \tag{3.16}$$

Clearly, the compressive gain decreases with increasing A_d. This gain compression is characterized by the signal level at which $A_d = A_{-1}$, i.e., at which the gain is 1 dB lower than the small signal gain, which is referred to as the *1-dB compression point* of the amplifier. The value of A_{-1} is found by setting,

$$G_c = 20 \log \left| a_1 \left(1 + \frac{3a_3}{4a_1} A_{-1}^3 \right) \right| = -1 \tag{3.17}$$

and solving for A_{-1}. This gives,

$$A_{-1} = \sqrt{\left(1 - 10^{-\frac{1}{20}} \right) \frac{4}{3} \left| \frac{a_1}{a_3} \right|} = \sqrt{0.145 \left| \frac{a_1}{a_3} \right|} \tag{3.18}$$

The concept of 1-dB compression point is illustrated in Fig. 3.8.

When there is interference present, the desired output signal will also decrease when the amplitude A_{I1} of the interferer increases, if $a_3 < 0$, see (3.15). In particular, the interference is responsible for a decrease in gain by the amount $3a_3 A_{I1}^2 / 2a_1$. This nonlinear effect is called *desensitization*, and manifests itself in that, when it occurs, the gain afforded a weak desired signal will be very small. In fact, gain decrease with the strong interference is *twice the rate* as compared to the gain compression case. Thus, if the interference is sufficiently strong, the gain will decrease to zero and the desired signal will be completely blocked.

3.3.2.2   Crossmodulation

When the input $x(t)$ consists of a weak desired signal $S_d(t)$ and a strong interferer $S_{I1}(t)$ with an amplitude modulation $1 + m(t)$, as follows:

$$\left. \begin{array}{l} S_d(t) = A_d \cos 2\pi f_0 t \\ S_{I1}(t) = A_{I1}\left[1 + m(t)\right]\cos 2\pi f_1 t \\ S_{I2}(t) = 0 \end{array} \right\} \tag{3.19}$$

The output, $y(t)$, of a nonlinear system is then:

$$y(t) = a_1 A_d \left[1 + \frac{3a_3}{4a_1} A_d^2 + \frac{3a_3}{2a_1} A_{I1}^2\left[1 + m^2(t) + 2m(t)\right]\right]\cos 2\pi f_0 t + \ldots \tag{3.20}$$

This expression contains desensitization and compression terms and two new terms, namely, $(3a_3 / 2a_1)A_{I1}^2 m^2(t)$, and $(3a_3 / a_1)A_{I1}^2 m(t)$. These new terms mean that the amplitude modulation on the strong interferer is now transferred to the desired signal through the interaction with the nonlinearity. This nonlinear phenomenon is referred to as *crossmodulation*.

3.3.2.3   Intermodulation

Consider the case when there is more than one interferer, $S_{I1}(t)$ and $S_{I2}(t)$ accompanying the desired signal $S_d(t)$ in the input $x(t)$,

$$\left. \begin{array}{l} S_d(t) = A_d \cos 2\pi f_0 t \\ S_{I1}(t) = A_{I1} \cos 2\pi f_1 t \\ S_{I2}(t) = A_{I2} \cos 2\pi f_2 t \end{array} \right\} \tag{3.21}$$

Then, the output becomes,

$$y(t) = a_1 A_d \left[1 + \frac{3a_3}{4a_1} A_d^2 + \frac{3a_3}{2a_2}\left(A_{I1}^2 + A_{I2}^2\right) \right] \cos 2\pi f_0 t$$

$$+ a_2 A_{I1} A_{I2} \left[\cos 2\pi\left(f_1 + f_2\right)t + \cos 2\pi\left(f_1 - f_2\right)t \right]$$

$$+ \frac{3}{4} a_3 \left[A_{I1}^2 A_{I2} \cos 2\pi\left(2f_1 \pm f_2\right)t + A_{I1} A_{I2}^2 \cos 2\pi\left(2f_2 \pm f_1\right)t \right] + \dots \quad (3.22)$$

In (3.22), the terms with frequencies $(f_1 + f_2)$, are called *second-order intermodulation products,* whereas the terms with frequencies $(2f_1 \pm f_2)$ and $(2f_2 \pm f_1)$, are called *third-order intermodulation products*. The latter frequencies are easily seen above in Fig. 3.6.

3.3.2.4   Memoryless Bandpass Nonlinearities

Power amplifiers (PAs) in wireless communications systems usually have a bandpass frequency response. These PAs are commonly considered memoryless, i.e., lacking storage elements, and, as a result, they can be represented by a nonlinear gain (AM-AM), and phase distortion, amplitude-to-phase conversion or AM-PM [3].

For a narrowband signal with carrier f_0, given by:

$$x(t) = A(t)\cos\left[2\pi f_0 t + \varphi(t)\right] \quad (3.23)$$

the PA response is given by:

$$y(t) = f[A(t)]\cos(2\pi f_0 t + \varphi(t) + g[A(t)]) \quad (3.24)$$

where $f[A(t)]$ is the nonlinear gain, *AM-AM conversion,* and $g(t)$ is the amplitude-to-phase *AM-PM conversion*. The typical magnitude and phase characteristic of a PA are shown in Fig. 3.9.

To be specific, AM/PM *Conversion* is defined as the slope of the phase shift vs. P_{in} or, equivalently, the amount of phase modulation caused by an input envelope variation of 1 dB, see Fig. 3.10. AM/PM *Transfer*, on the other hand, is the equivalent phase modulation induced on an unmodulated carrier by another carrier with 1 dB of amplitude modulation [3].

The typical amplifier model for system-level analysis is given in Fig. 3.11.

3.3.3   *Mixers*

The mixer is a nonlinear device employed to effect frequency translation, e.g., up- or down-conversion. When two inputs are applied to a device with a square-law current-voltage curvature, the output contains *sum and difference* frequencies of the two input tones.

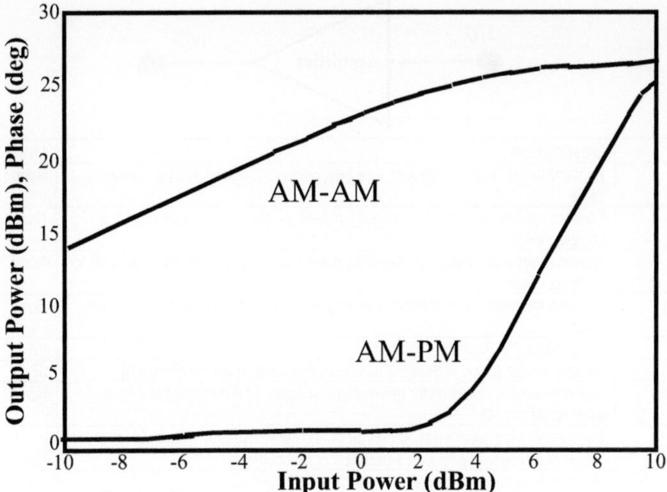

Fig. 3.9 Characterization of Class A-B power amplifier

Fig. 3.10 AM/PM conversion

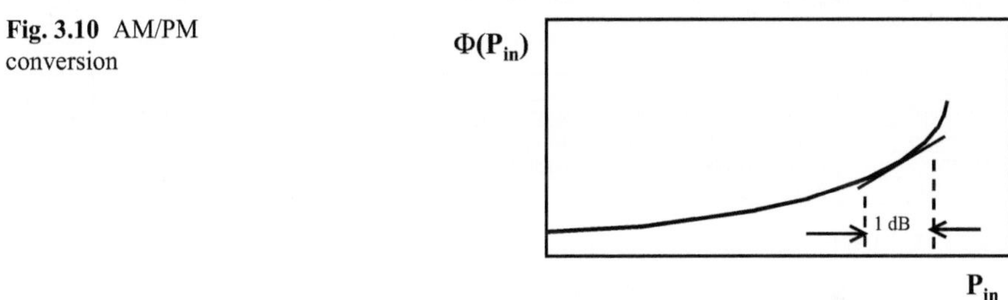

Mixer performance is captured by the following parameters:

1. Conversion Gain (Loss): Is the ratio of the output (IF) signal power to the input (RF) signal power.
2. Noise Figure: Is the SNR at the input (RF) port divided by the SNR at the output (IF) port.
3. Isolation: Represents the amount of "leakage" or "freedthrough" between the mixer ports.
4. Conversion Compression: Relates to the RF input power level above which the curve of IF output power versus RF input level deviates from linearity.
5. Dynamic Range: Is the amplitude range over which the mixer can operate without performance degradation. It is dictated by the conversion compression and the noise figure.
6. Two-Tone, Third-Order Intermodulation Distorsion: Is the amount of third-order distorsion caused by the presence of a second received signal at the RF port.
7. Intercept Point: Is the point at which the fundamental response and the third-order sputious response curves intersect.

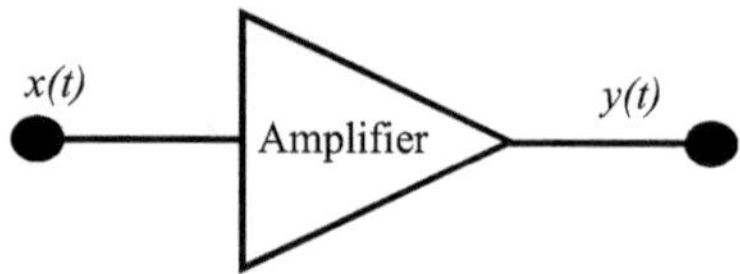

Name	Description	Units	Default
S21	Forward transmission coefficient, use x+jy, polar(x,y), dbpolar(x,y) for complex value	None	dbpolar(0,0)
S11	Forward reflection coefficient, use x+jy, polar(x,y), dbpolar(x,y),vswrpolar(x,y)) for complex value	None	polar(0,0)
S22	Reverse reflection coefficient, use x+jy, polar(x,y), dbpolar(x,y),vswrpolar(x,y)) for complex value	None	polar(0,180)
S12	Reverse transmission coefficient, use x+jy, polar(x,y), dbpolar(x,y) for complex value	None	0
NF	Noise figure [NF mode for NFmin=0]	dB	None
NFmin	Minimum noise figure at Sopt[(NFmin, Sopt,Rn) mode used for NFmin>0]	dB	None
Sopt	Optimum source reflection for minimum noise figure [(NFmin, Sopt,Rn) mode used for NFmin>0]	None	None
Rn	Equivalent noise resistance[(NFmin, Sopt,Rn) mode used for NFmin>0]		
Z1	Reference impedance for port 1 (must be a real number)	None	
Z2	Reference impedance for port 2 (must be a real number)	None	
GainCompType	Gain compression type	None	List
GainCompFreq	Frequency at which Gain Compression is specified	None	
ReferToInput	Specify each of the gain compression options notified in Polynomial Order for various Magnitude Modes with respect to the input or output power of the device	None	Ooutput
SOI	Second order intercept	dBm	None
TOI	Third order intercept	dBm	None
Psat	Power saturation point (always referred to input, regardless of the value of the ReferToInput parameter)	dBm	None
GainCompSat	Gain compression at Psat	dB	5
GainComPower	Power level at gain compression specified by GainComp	dBm	None
GainComp	Gain Compression to phase modulation	dB	1
AM2PM	Amplitude modulation to phase modulation	deg/dB	None
PAM2PM	Power level at AM2PM	dBm	None

Fig. 3.11 Parameters describing amplifier model

8. Desensitization: Is the compression at the desired signal frequency caused by a strong interfering signal on an adjacent frequency.
9. Harmonic Intermodulation Distortion: Results from the mixing of mixer-generated harmonics of the input signals. These distortion products have frequencies where m and n represent the harmonic order.
10. Cross-Modulation Distorsion: Is the amount of modulation transferred from a modulated carrier to an unmodulated carrier when both signals are applied to the RF port.

The terminology of mixer performance is illustrated in Fig. 3.12.

A problem with all mixer circuits is their *spurious response*. This is the production of outputs at the intermediate frequency f_{IF} due to signals at frequencies *other than* the desired received frequency f_{RF}. These other frequency components may be: (1) Coming directly from the antenna, if no RF preamplifier stage is used; (2) Produced by nonlinear action in the RF amplifier; (3) Produced in the mixer itself; (4) Attributable to oscillator-frequency harmonics. Figure 3.13 illustrates these spurious frequency components.

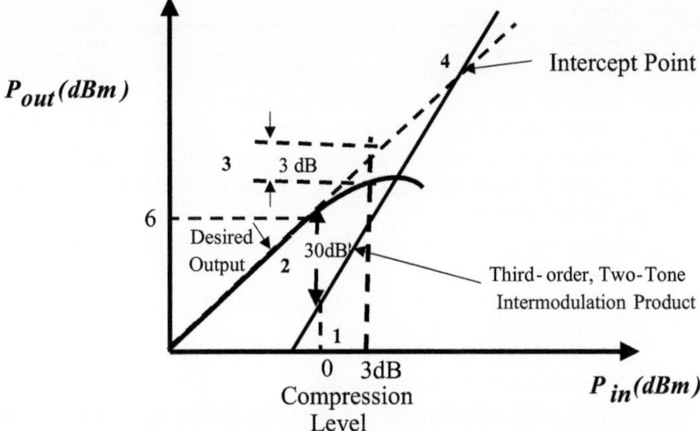

Fig. 3.12 Transfer characteristic of mixer's IF output power vs. input power. The *bold* numbers indicate the following properties: *1*. At 0 dBm input, the output is 6 dBm, indicating a 6 dB conversion gain. *2*. At this input level the *3rd-order, two-tone intermodulation product* is 30 dB below the desired output. *3*. At a higher input value, the 3 dB compression point is indicated. *4*. At still higher input level, the intercept point is shown where the projected curves of desired output and *3rd-order intermodulation product* intersect

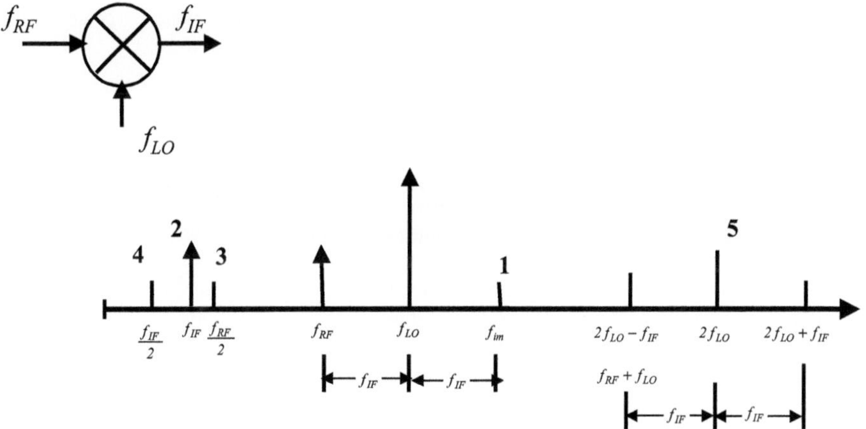

Fig. 3.13 Sprious frequency products in mixers. The *bold* numbers indicate the following: *1*.The image frequency $f_{im}=f_{LO}+f_{IF}$. If a signal at this frequency is picked up by the antenna, and if it reaches the mixer input, it will beat with f_{LO} to produce a difference-frequency component equal to f_{IF}. *2*. An input signal at f_{IF} will appear in the output due to normal amplifier action. *3*. An input frequency equal to $f_{RF}/2$ may be doubled to f_{RF} by the square-law mixer term and then combine with f_{LO} to produce output at f_{IF}. *4*. An input at $f_{IF}/2$ may be doubled by the mixer and appear in the output. *5*. If the LO output includes a second harmonic at $2f_{LO}$, or if the mixer generates $2f_{LO}$, this component can beat with received inputs at $(2f_{LO} \pm f_{IF})$ to produce output at f_{IF}.

The fact that many possibilities exist, as indicated in Fig. 3.12, illustrates the need for adequate selectivity in front of the mixer stage, and for good linearity in

Table 3.1 Mixer Model

Name	Description	Units	Default
StyleSideband	Specify the sideband/image option for the mixer	None	BOTH
OutputSideSuppression	Output sideband suppression (only relevant for sideband = LOWER, UPPER)	dB	-200
InputImageRejection	Input image rejection (only relevant for sideband = LOWER IMAGE REJECTION, UPPER IMAGE REJECTION)	dB	-200
ConvGain	Conversion gain; use x+jy, polar(x, y), dbpolar(x, y) for complex	None	dbpolar(0, 0)
RevConvGain	Reverse conversion gain, use x+jy, polar(x, y), dbpolar(x, y) for complex value	None	polar(0, 0)
SP11	S11, RF port reflection, polar(x, y), dbpolar(x, y), vswrpolar(x, y) for complex data	None	polar(0,0)
SP12	S12, IF port to RF port leakage, use x+jy, polar(x, y), dbpolar(x, y), vswrpolar(x, y) for complex data	None	polar(0, 0)
SP13	S13, LO port to RF port leakage, use x+jy, polar(x, y), dbpolar(x, y), vswrpolar(x, y) for complex data	None	polar(0,0)
SP21	S21, RF port to IF port leakage, use x+jy, polar(x, y), dbpolar(x, y), vswrpolar(x, y) for complex data	None	polar(0, 0)
SP22	S22, IF port reflection, use x+jy, polar(x, y), dbpolar(x, y), vswrpolar(x, y) for complex data	None	polar(0, 0)
SP23	S23, LO to IF port leakage, use x+jy, polar(x, y), dbpolar(x, y), vswrpolar(x, y) for complex data	None	polar(0, 0)
SP31	RF port to LO leakage (real or complex number)S31, RF port to LO port leakage, use x+jy, polar(x, y), dbpolar(x, y), vswrpolar(x, y) for complex value	None	polar(0, 0)
SP32	IF port to LO leakage (real S32, IF port to LO leakage, use x+jy, polar(x, y), dbpolar(x, y), vswrpolar(x, y) for complex value or complex number	None	polar(0, 0)
SP33	LO port reflection (real or complex num S33, LO port reflection, use x+jy, polar(x, y), dbpolar(x, y), vswrpolar(x, y) for complex value or complex number	None	polar(0, 0)
PminLO	Minimum LO power before starvation	dbM	-100
DetBW	Detector bandwidth for LO limiting	Hz	1E100
NF	Double sideband noise figure	dB	None
NFmin	Minimum double sideband noise figure at Sopt	dB	None
Sopt	Optimum source reflection for minimum noise figure, use x+jy, polar(x, y), dbpolar(x, y) for complex value	None	None
Rn	Equivalent noise resistance		
Z1	Reference impedance for port 1 (must be a real number)	50	
Z2	Reference impedance for port 2 (must be a real number)	50	
Z3	Reference impedance for port 3 (must be a real number)	50	
GainCompType	Gain compression type	None	LIST
GainCompFreq	Reference frequency for gain compression if gain compression is described as a function of frequency		

Table 3.1 (continued)

Name	Description	Units	Default
ReferToInput	Specify gain compression with respect to input or output power of device	None	OUTPUT
SOI	Second order intercept	dBm	None
TOI	Third order intercept	dBm	None
Psat	Power saturation point (always referred to input, regardless of the value of the ReferToInput parameter)	dBm	None
GainCompSat	Gain compression at Psat	dB	5.0
GainComp- Power	Power lebel in dBm at gain compression for X dB compression point, specified by GainComp	dBm	None

the RF stage, to avoid the generation of spurious frequencies at that point. A set of parameters capturing the mixer behavior is given in Table 3.1.

3.3.4   Filters

Filters provide frequency selection by allowing the transmission of some frequencies and attenuating that of others. In addition to minimum attenuation in the passband, it may require linear phase (constant delay) across the passband. Ideally, a filter's response would exhibit infinitely sharp amplitude transitions from passband to stopband, Fig. 3.14a, however, in reality this is not possible, and the ideal response is only approximated, Fig. 3.14b.

A typical filter model, employed in systems analyses, is shown in Fig. 3.15.

3.3.5   Oscillators

The oscillator is a circuit that generates a stable output power and frequency (e.g., a carrier). Its performance is characterized by the degree of stability of the frequency, called *phase noise*, and of its output signal amplitude, called *amplitude noise*. The general oscillator waveform, including these types of noise, may be expressed as,

$$A_{Osc}(t) = (A + a_n(t))\cos(2\pi f_0 t + \varphi_n(t)) \tag{3.25}$$

where $a_n(t)$ and $\phi_n(t)$ capture the origin of the amplitude and phase noise, respectively.

3.3.5.1   Amplitude Noise

Clearly, if $\phi_n(t) = 0$, in (3.25), the oscillator output signal will obey the expression,

Fig. 3.14 a Ideal and **b** real filter responses

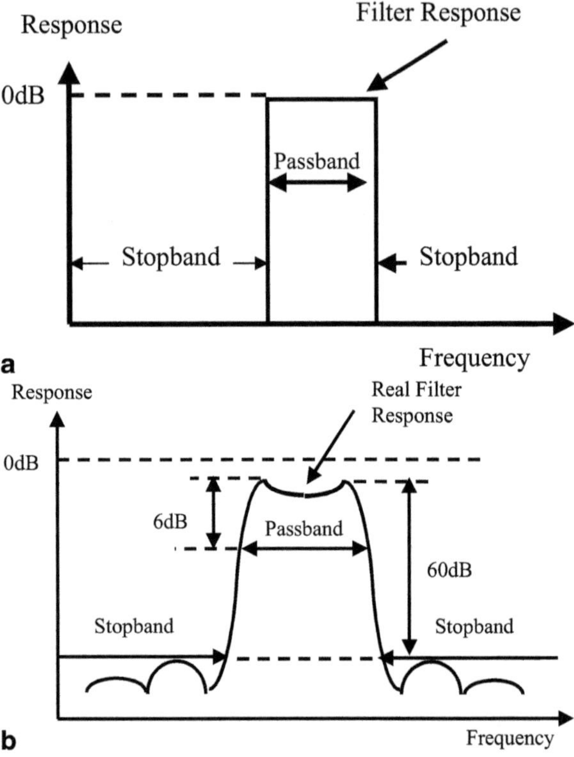

Name	Description	Units	Default
Fcenter	Passband center frequency	GHz	1.5
BWpass	Passband edge-to-edge width	GHz	1.0
Ripple	Passband ripple	dB	1
BWstop	Stopband edge-to-edge width	GHz	1.2
Astop	Attenuation at stopband edge	dB	20
StopType	Stopband input impedance type: OPEN or SHORT	None	Open
BWstop	Stopband edge-to-edge width	GHz	1.2
MaxRej	Maximum rejection level	dB	None
N	Filter Order (if N>0, it overwrites GDpass	None	0
IL	Passband insertion loss	dB	0
Qu	Unloaded quality factor for resonators, default setting is an infinite Qu and expresses a dissipationless resonant circuit	None	1e308
Z1	Input port reference impedance	Ohm	50
Z2	Output port reference impedance	Ohm	50
Temp	Temperature	°C	None
Z2	Output port reference impedance	Ohm	50

Fig. 3.15 Typical bandpass filter model

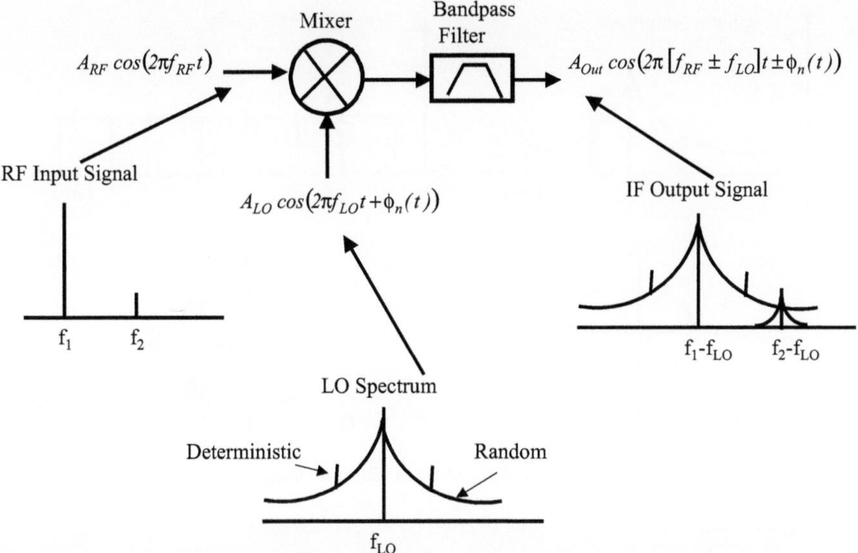

Fig. 3.16 Impact of local oscillator (*LO*) noise on the noise of a down-converted *RF* signal

$$A'_{Osc}(t) = (A + a_n(t))\cos(2\pi f_0 t) \tag{3.26}$$

which is the equation of an amplitude modulated signal. As studied previously, the output spectrum will then include, in addition to f_0, two sidebands (in the narrowband case). On the other hand, if $a_n(t) = 0$, the oscillator output signal will obey the expression,

$$A''_{Osc}(t) = A\cos(2\pi f_0 t + \varphi_n(t)) \tag{3.27}$$

which resembles a frequency modulated signal. As studied previously, the output spectrum will then be broadened around the desired frequency, f_0. This is visualized next in the context of a local oscillator (LO).

3.3.5.2 Phase Noise of a Local Oscillator

The phase modulation of the local oscillator is transferred to the incoming signal, see Fig. 3.16 [3]. In particular, the modulation $\phi_n(t)$ can be: (1) Deterministic (δ functions in frequency domain), or (2) Random (continuous spectrum).

The phase noise in an oscillator is manifested in both the frequency and the time domains. In the former, Fig. 3.17a, it appears as a broadening in the amplitude of the frequency of interest, f_0, i.e., the output frequency "jitters" around f_0. In the latter, Fig. 3.17b, it appears as jitter in the time position of a digital pulse (clock). In addition to these, there is a practical impact of phase noise on the actual signals transmitted. For instance, there is a displacement of the constellations, Fig. 3.17c, d, so that their EVM increases [3].

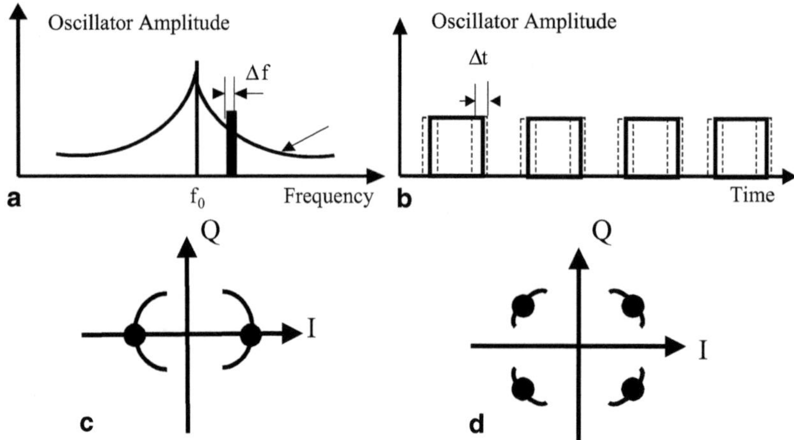

Fig. 3.17 Manifestations of oscillator phase noise. **a** *Oscillator amplitude* in the *frequency* domain (frequency spectrum), **b** *Oscillator amplitude* in the *time* domain, **c** Impact of timing jitter on BPSK constellation, **d** Impact of timing jitter on QPSK constellation

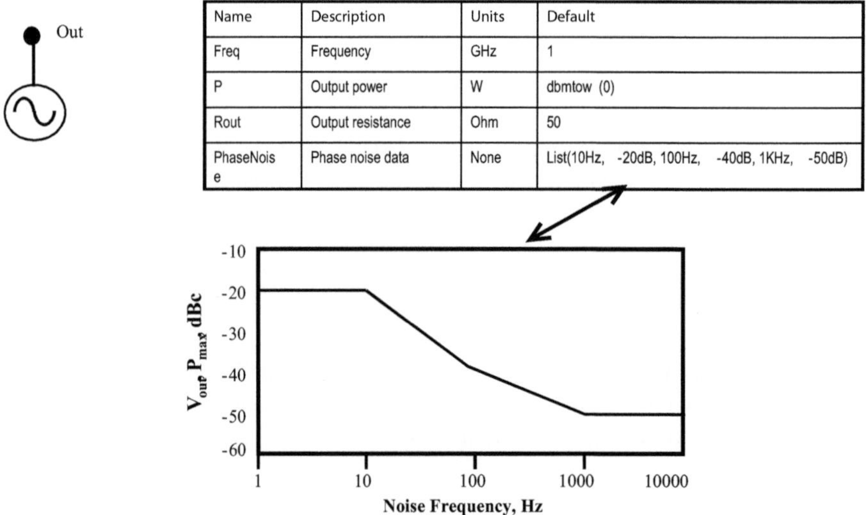

Name	Description	Units	Default
Freq	Frequency	GHz	1
P	Output power	W	dbmtow (0)
Rout	Output resistance	Ohm	50
PhaseNoise	Phase noise data	None	List(10Hz, -20dB, 100Hz, -40dB, 1KHz, -50dB)

Fig. 3.18 Oscillator model

A typical oscillator model, Fig. 3.18, includes its frequency, output power and phase noise properties.

3.3.6 *Frequency Multipliers*

Frequency multipliers (FM) are circuits whose output signal is a harmonic of the input signal. Many aspects of FMs are common to power amplifiers, in particu-

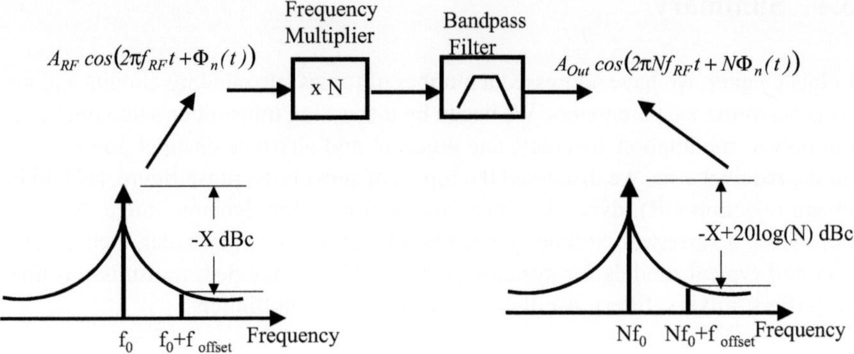

Fig. 3.19 Phase noise enhancement by frequency multipliers

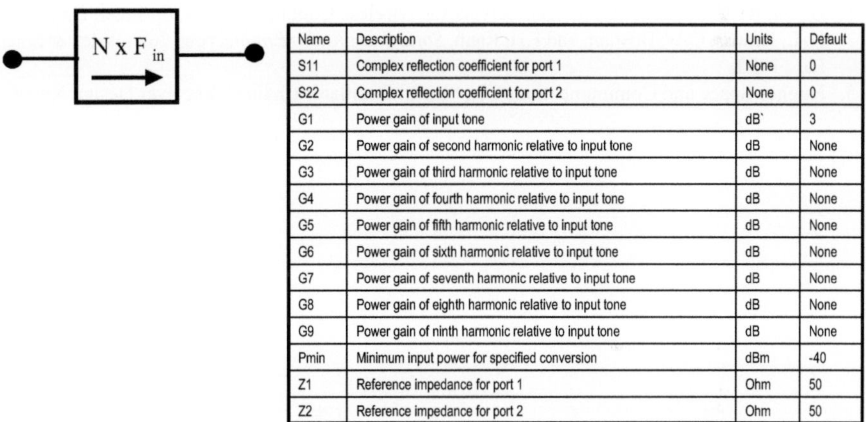

Name	Description	Units	Default
S11	Complex reflection coefficient for port 1	None	0
S22	Complex reflection coefficient for port 2	None	0
G1	Power gain of input tone	dB`	3
G2	Power gain of second harmonic relative to input tone	dB	None
G3	Power gain of third harmonic relative to input tone	dB	None
G4	Power gain of fourth harmonic relative to input tone	dB	None
G5	Power gain of fifth harmonic relative to input tone	dB	None
G6	Power gain of sixth harmonic relative to input tone	dB	None
G7	Power gain of seventh harmonic relative to input tone	dB	None
G8	Power gain of eighth harmonic relative to input tone	dB	None
G9	Power gain of ninth harmonic relative to input tone	dB	None
Pmin	Minimum input power for specified conversion	dBm	-40
Z1	Reference impedance for port 1	Ohm	50
Z2	Reference impedance for port 2	Ohm	50

Fig. 3.20 Frequency multiplier model

lar, in both the generation of harmonic signals derives from a nonlinear transfer characteristic. The amplitude of the harmonics depends upon both the amplitude of the driving signal and the abruptness of the nonlinearity. An important aspect of FMs has to do with their contribution to the increase in phase noise. In particular, if a random noise $\Phi_n(t)$ is added to the input FM signal, then this phase noise will appear at the output multiplied by N, the FM multiplication factor. As a result, treating sidebands as phase modulation of the carrier frequency by $\Phi_n(t)$, the effective modulation index M_p is increased by N, and the sidebands are increased by 20log(N) dB, Fig. 3.19 [3].

A typical model for frequency multipliers used in systems analysis is shown in Fig. 3.20.

3.4 Summary

In this chapter, we have addressed a number of topics surrounding circuits and systems performance parameters. We began by discussing transmitter's maximum output power, modulation accuracy, and adjacent and alternate channel power. Then, on the receiver side, we discussed the topics of sensitivity, noise figure, selectivity, image rejection (IR), dynamic range, and spurious free dynamic range. Next, we introduced a variety of circuits/systems building blocks, in particular, their description and typical models for systems analysis. These included, transmission lines, amplifiers, mixers, filters, oscillators, and frequency multipliers.

References

1. Q. Gu, RF System Design of Transceivers for Wireless Communications, Springer, 2005.
2. H. L. Krauss, C.W. Bostian, and F.H. Raab, *Solid State Radio Engineering*, John Wiley & Sons, Inc., 1980.
3. Hughes Space and Communications Co., "Local Oscillator Chains," Receiver Design Notes.

Chapter 4
Radio Channel Fundamentals and Antennas

Abstract In this chapter, we begin by discussing the fundamentals of the radio channel. Typical Radio Transmitter and Receiver block diagrams are reviewed, together with the components of a wireless communications system. Then, we turn to addressing the phenomenon of wave propagation, in particular, Multi-Path Propagation, Free-Space, and Path Loss. The relationship between power and electric field is elucidated, the Effective Isotropic Radiated Power is defined, and the measured received electric field is related to the received power. Then, the various models invoked in calculating the effects of propagation are presented. These include: Reflection, in particular, the conditions of: (1) Two Half-Spaces: Dielectrics; (2) Two Half-Spaces: Metals; (3) Two Half-Spaces: Multi-Layer Structures; and (4) Two-Ray Ground Model. This is followed by a discussion of Diffraction, in particular, from a Knife-Edge. We conclude this discussion with the topics of Scattering, and Reflection in the Troposphere. The chapter is concluded with a discussion of antenna parameters, in particular: (1) Efficiency; (2) Effective Area; (3) Gain; and (4) Directivity.

4.1 Wireless Radio Channel

We began this book by giving an overview of the wireless communications problem, Fig. 4.1. In particular, the elements comprising the transmitter and the receiver were discussed in Chap. 1 through 3. In this Chapter, we focus on a variety of aspects that affect the signal as it propagates between transmitter and receiver [1].

4.1.1 Radio Transmitter and Receiver Block Diagram

A more functional depiction of modern wireless communication systems is given by the block diagram in Fig. 4.2 [2]. It is seen that, on the transmitter side, analog and digital signal sources are first processed by analog-to-digital converters, source encoders, and channel encoders. Then, they undergo pulse shaping, filtering,

H. J. De Los Santos et al., *Radio Systems Engineering,*
DOI 10.1007/978-3-319-07326-2_4, © The Authors 2015

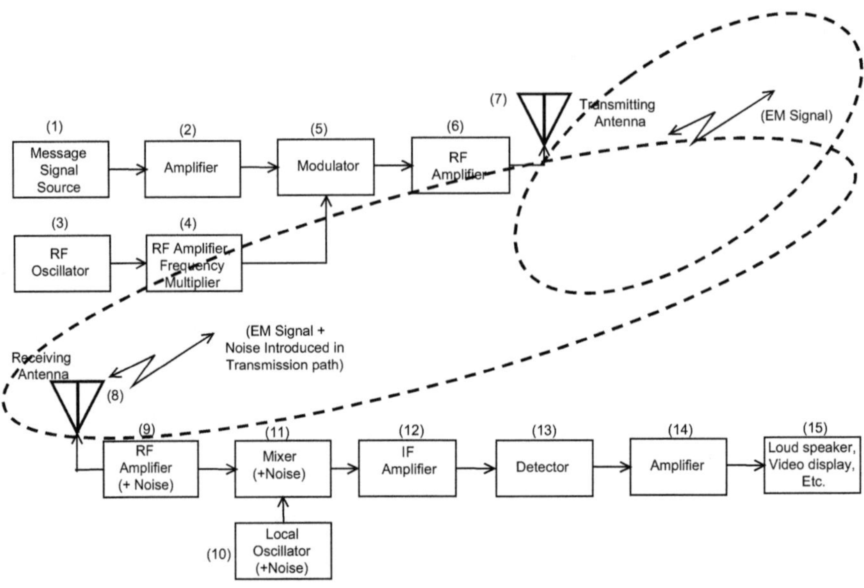

Fig. 4.1 Diagram that captures the wireless communications problem

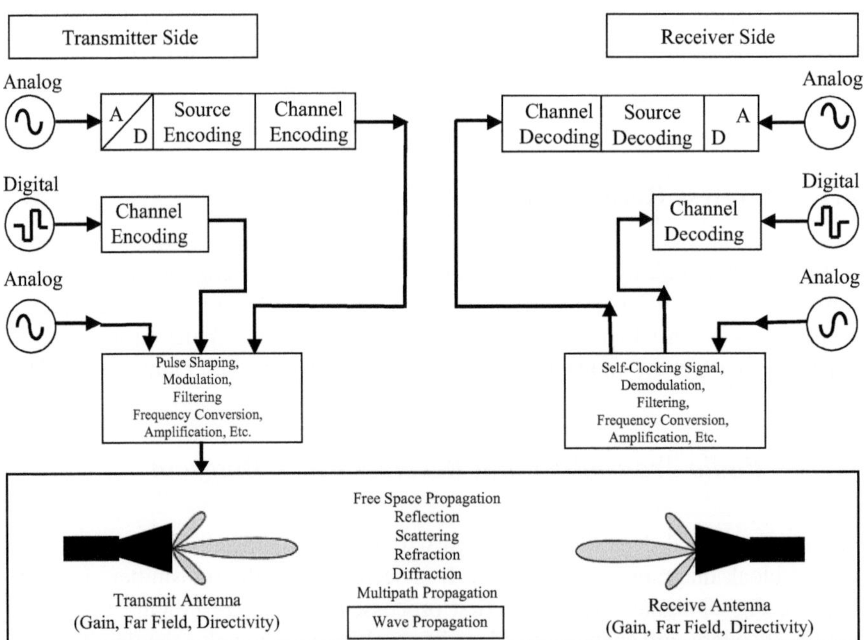

Fig. 4.2 Components of a modern wireless communication system, including antenna and wave propagation medium parameters. *After* [2]

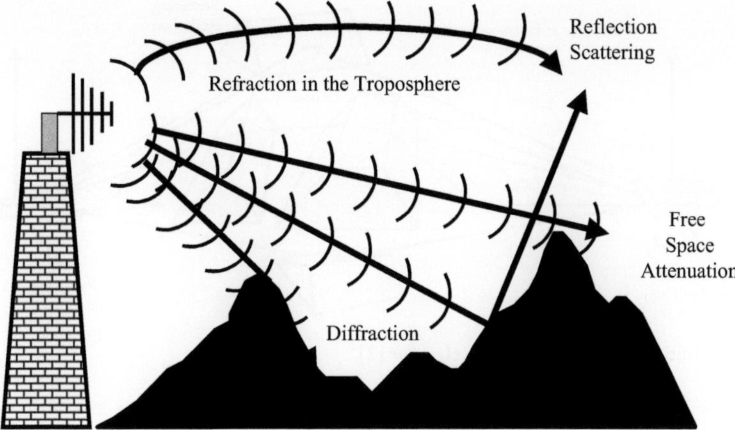

Fig. 4.3 Sources of attenuation as a wave propagates from source to destination. *After* [2]

frequency conversion, and amplification, as well as used to effect modulation, before being applied to the transmit antenna. The nature of the transmit antenna, in particular, its gain, far field and directivity properties, determine the final signal features being transmitted. Once the signal has been launched into free space, its features are impacted by the properties of the propagation medium, which include, reflection, scattering, refraction, diffraction, and multipath propagation. Finally, on the receiver side, the signal features are again impacted by the gain, far field, and directivity properties of the receiving antenna, whose output feeds the receiver front-end where the function of self-clocking, demodulation, filtering, frequency conversion, amplification, etc., are effected prior to decoding the baseband information.

4.1.2 Wave Propagation Phenomena

En-route from the transmitter to the receiver, the propagating signal wave experiences a variety of physical effects that contribute to decreasing the power received at the destination. With reference to Fig. 4.3, these include (as per the respective numbers):

1. Free Space Attenuation: This is caused by the natural decay with distance from the Source.
2. Reflection Scattering: This is caused by the surface of the Earth, Buildings and Walls with dimensions much greater than the wavelength.
3. Diffraction: This is caused by surfaces with sharp irregularities (Edges).
4. Reflection Scattering: This is caused by objects with dimensions much smaller than the wavelength, and when the number of obstacles per unit volume is large.
5. Refraction in the troposphere.

These various means of signal attenuation are discussed in more detail next.

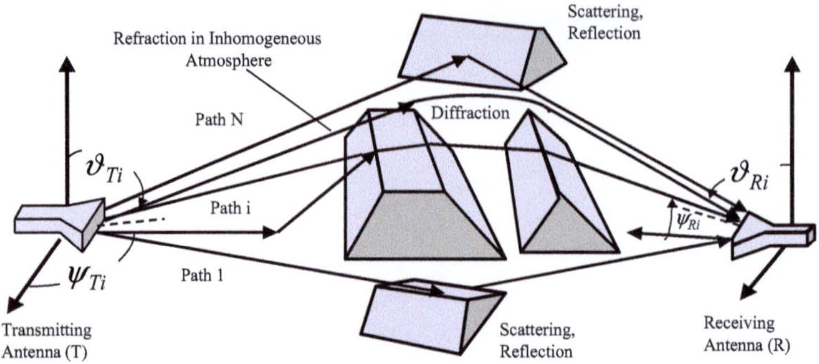

Fig. 4.4 Multi-path propagation model. *After* [2]

4.1.3   Multi-Path Propagation

The multipath propagation problem, Fig. 4.4, deals with finding the answer to the question: What is the received power as a result of propagation through ALL paths? Answering this question pertains to the realm of *Channel Modeling*, which involves the combination of analytical, numerical, statistical, and specific measurements, in other words, it is a difficult problem. Nevertheless, a number of models have been developed to quantify/approximate the result, and these will be discussed next.

4.1.3.1   Free Space Propagation Model

The model for free space propagation, Fig. 4.4, is used to predict the strength of the signal received when the transmitter and receiver have a clear, *unobstructed line-of-sight path* between them [3]. This situation is manifested when dealing with propagation in satellite communications and microwave line-of-sight radio links. The free space model predicts that the received power decays as a function of the separation between transmitter and receiver raised to some power [3]. This is captured by the *Friis* free space equation [3]:

$$P_r(d) = \frac{P_t G_t G_r \lambda^2}{(4\pi)^2 d^2 L} \tag{4.1}$$

where P_t is the transmitted power, $P_r(d)$ is the received power, d is the transmitter-receiver (T-R) separation distance in meters, G_t is the transmitter antenna gain, G_r is the receiver antenna gain, L is the system loss factor not related to propagation, and λ is the wavelength in meters. The antenna gain is related to its effective aperture, A_e, by,

$$G = \frac{4\pi A_e}{\lambda^2} \tag{4.2}$$

where λ is related to the carrier frequency by,

$$\lambda = \frac{c}{f} = \frac{2\pi c}{\omega_c}$$ (4.3)

f is in Hertz, ω_c is in radians per second, and c is the speed of light in meters/s. Clearly, P_t and P_r must be expressed in the same units, and G_t and G_r are dimensionless quantities. Finally, L represents miscellaneous losses, e.g., filter losses; a value of $L=1$ indicates that there is no loss in the system hardware. The Friis formula, (4.1), shows that the received power falls off as the square of the T-R separation distance. This implies that the received power decays with distance at a rate of 20 dB/decade [3].

A parameter that captures the power loss between transmitter and receiver, utilized in overall transmission analysis ("link") calculations, is the so-called *path loss*. This is the difference between the effective transmitted power and the received power, expressed in dB [3],

$$PL(dB) = 10\log\frac{P_t}{P_r}$$ (4.4)

When the antenna gains are included, the path loss is given by,

$$PL(dB) = -10\log\left[\frac{G_t G_r \lambda^2}{(4\pi)^2 d^2}\right]$$ (4.5)

but, on the other hand, when antenna gains are excluded, they are assumed to have unity gain, and the path loss is given by [3],

$$PL(dB) = -10\log\left[\frac{\lambda^2}{(4\pi)^2 d^2}\right]$$ (4.6)

The model described by Friis' formula is only valid under certain circumstances, namely, for values of d in the *far field* of the transmitting antenna. The far field region, in turn, is defined by,

$$d_f = \frac{2D^2}{\lambda}$$ (4.7)

where D is the largest linear dimension of the antenna, and $d_f \gg D$ and $d_f \sim \lambda$. Since $P_r(d)$ is undefined for $d=0$, a reference point d_0 is employed for large-scale propagation models. Then, the received power at distances $d>d_0$, may be related to that at d_0, and this is expressed as [3],

$$P_r(d) = P_r(d_0)\left(\frac{d_0}{d}\right)^2 \quad d \geq d_0 \geq d_f$$ (4.8)

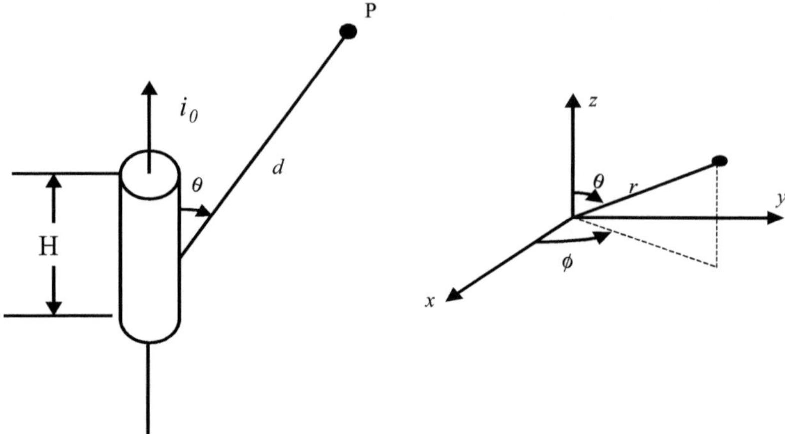

Fig. 4.5 Linear electric field radiator

In the case of mobile radio systems, it is found that the received power may change by many orders of magnitude over a typical coverage area of several km^2, therefore, it has been to be more convenient to express it in dBm [3],

$$P_r(d)dBm = 10\log\left[\frac{P_r(d_0)}{0.001W}\right] + 20\log\left(\frac{d_0}{d}\right) \quad d \geq d_0 \geq d_f \qquad (4.9)$$

where $P_r(d_0)$ is in units of Watts. Typical values for d_0 are: *1 m* in indoor environments, *100 m* or *1 km* in outdoor environments [3].

As is well known, the radiated power is distributed in space as dictated by the antenna properties, therefore, a relationship between power and field is necessary. A fundamental expression for this relationship is obtained by considering a linear radiator of extent H ($H \ll \lambda$), carrying a current of amplitude i_0 and making an angle θ with a point P, at a distance d, Fig. 4.5.

It can be shown, in fact, that such a linear radiator produces the following fields [3],

$$E_r = \frac{i_0 L \cos\theta}{2\pi\varepsilon_0 c}\left\{\frac{1}{d^2} + \frac{c}{j\omega_c d^3}\right\}e^{j\omega_c(t-d/c)} \qquad (4.10)$$

$$E_\theta = \frac{i_0 L \cos\theta}{4\pi\varepsilon_0 c^2}\left\{\frac{j\omega_c}{d} + \frac{c}{d^2} + \frac{c^2}{j\omega_c d^3}\right\}e^{-j\omega_c(t-d/c)} \qquad (4.11)$$

$$H_\phi = \frac{i_0 L \cos\theta}{4\pi c}\left\{\frac{j\omega_c}{d} + \frac{c}{d^2}\right\}e^{j\omega_c(t-d/c)} \qquad (4.12)$$

Fig. 4.6 Sphere depicting
radiating source and unit
area on a sphere of radius d,
receiving the power density

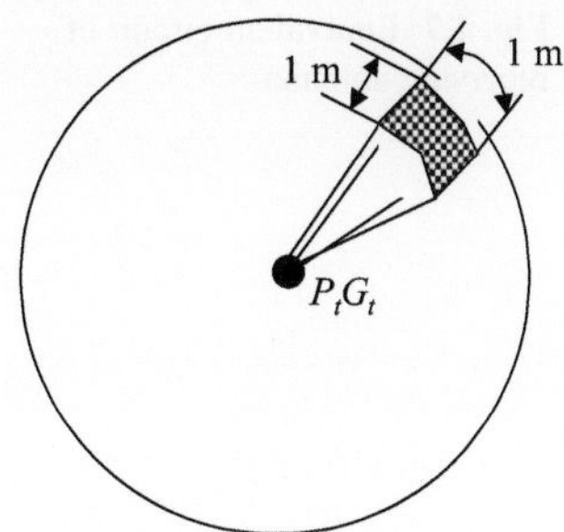

where the $1/d$ terms represent *radiation field* components, the $1/d^2$ terms represent *induction field* components, and the $1/d^3$ terms represent *electrostatic field* components. At distances far from the antenna, only non-negative radiation fields with the $1/d$ dependence remain [2–3].

The total power radiated by an antenna is the given by the product of the transmitted power and the transmitter antenna gain, P_tG_t, called the *Effective Isotropic Radiated Power* (EIRP). Due to its (up to this point) isotropic nature, however, the amount of radiated power reaching a distance d, away from the antenna, is given by the product of the radiated power *density*, given by,

$$P_d = \frac{EIRP}{4\pi d^2} = \frac{P_tG_t}{4\pi d^2} = \frac{E^2}{R_{fs}} = \frac{E^2}{\eta} \ W/m^2 \tag{4.13}$$

and the effective area of the receiving antenna; R_{fs} is the intrinsic impedance of free space, $\eta = 120\pi \ \Omega$ (377 Ω), Fig. 4.6 [3].

Using this value of R_{fs}, we have,

$$P_d = \frac{|E|^2}{377\Omega} \ W/m^2 \tag{4.14}$$

For an effective aperture area of the receiving antenna, A_e, the received power is,

$$P_r(d) = P_d A_e = \frac{|E|^2}{120\pi}\cdot A_e = \frac{P_tG_tG_r\lambda^2}{(4\pi)^2 d^2} = \frac{|E|^2 G_r\lambda^2}{480\pi^2} \ W \tag{4.15}$$

The voltage detected upon receiving the power P_r at a distance d from the transmitter is given by,

$$P_r(d) = \frac{V^2}{R_{ant}} = \frac{V_{ant}/2^2}{R_{ant}} = \frac{V_{ant}^2}{4R_{ant}} \tag{4.16}$$

This equation derives from the equivalent circuit of the receiving antenna, Fig. 4.7, characterized by its open circuit voltage, and its radiation resistance [3]. For maximum detection, the antenna output voltage, V, must be matched to the receiver circuit.

Fig. 4.7 Equivalent circuit of
receiving antenna

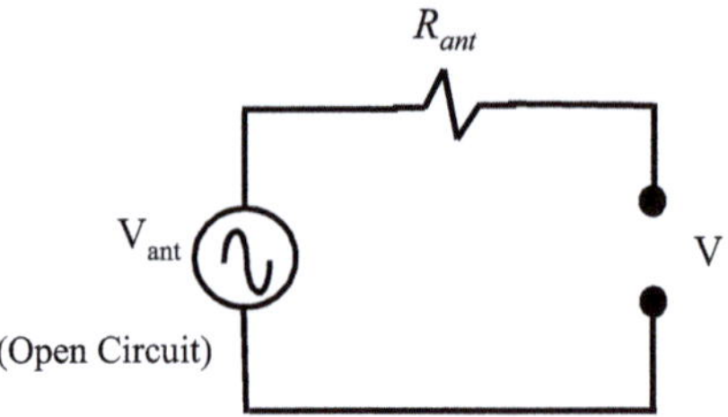

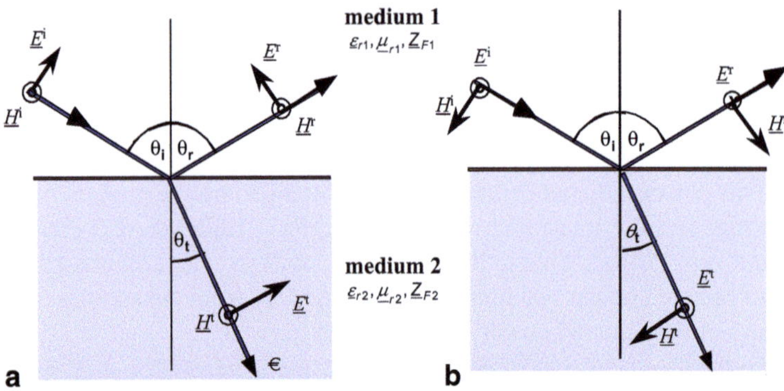

Fig. 4.8 Relations between incident electric field and reflecting media. **a** E field parallel to plane
of incidence, *TM polarization*. **b** E field orthogonal to plane of incidence, *TE polarization*

4.1.3.2 Reflection Models

As the signal propagates from transmitter to receiver, the field resulting from reflection at a variety of materials and configurations is computed by solving Maxwell's equations. This gives rise to three reflection models, corresponding to whether the materials are dielectrics, perfect conductors, or ground [2]. The models depend on whether the electric (E) field is parallel to the plane of incidence, in which case the field is said to be in the *TM polarization*, or whether the E field is orthogonal to plane of incidence, in which case it is said to be in the *TE polarization*, Fig. 4.8.

In the TM case, the reflection coefficient for wave reflection between a medium 1 and a medium 2, occupying two half spaces, is captured in Fig. 4.9a.

The reflected and transmitted fields are given in terms of the so-called *Fresnel* reflection (4.17) and transmission (4.18) coefficients, as follows:

$$\vec{E}' = R_{\parallel}\vec{E}^i, \quad R_{\parallel} = \frac{Z_{F1}\cos\theta_i - Z_{F2}\cos\theta_t}{Z_{F1}\cos\theta_i + Z_{F2}\cos\theta_t} \tag{4.17}$$

$$\vec{E}^t = T_{\parallel}\vec{E}^i, \quad T_{\parallel} = \frac{2Z_{F2}\cos\theta_i}{Z_{F1}\cos\theta_i + Z_{F2}\cos\theta_t} \tag{4.18}$$

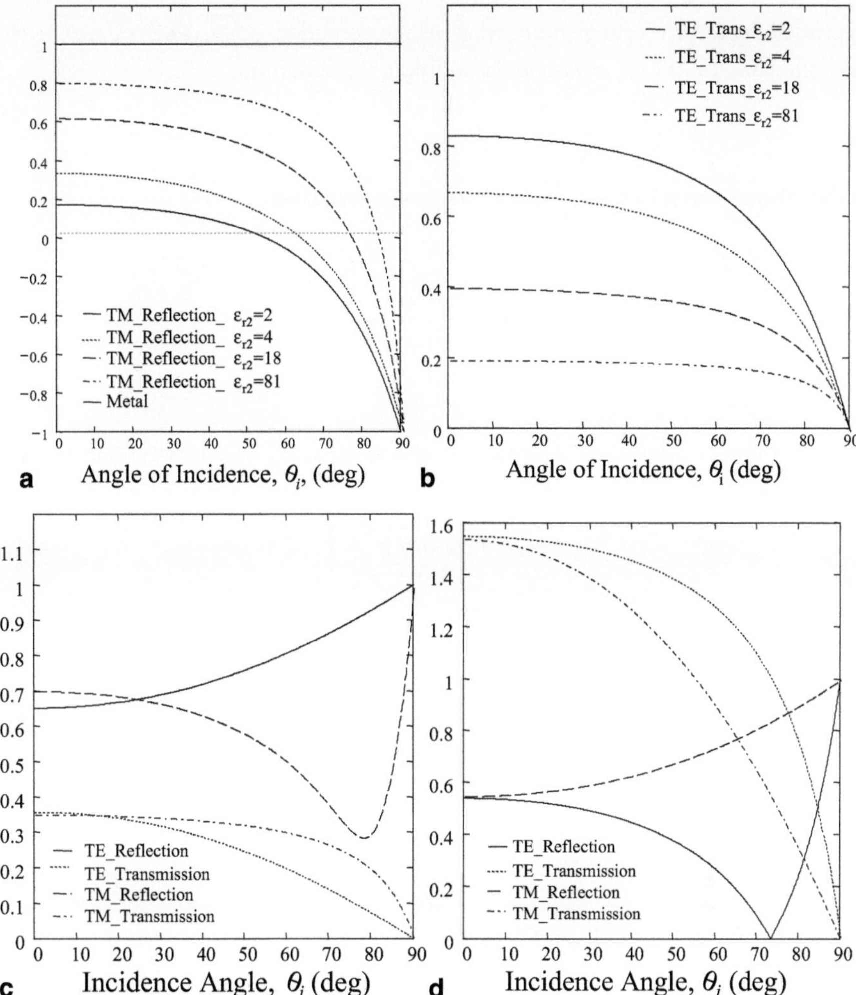

Fig. 4.9 Dependence of various quantities with respect to the angle of incidence. **a** Reflection coefficient $R_{//}$ for parallel (TM) polarization, for $\mu_{r1} = \mu_{r2} = 1$, $\varepsilon_{r1} = 1$, $\varepsilon_{r2} = 2$. **b** Reflection coefficient $R_{\perp}$ for parallel polarization for $\mu_{r1} = \mu_{r2} = 1$, $\varepsilon_{r1} = 1$, $\varepsilon_{r2} = 2$. Reflection and transmission coefficient behavior for various permittivities. **c** $\mu_{r1} = \mu_{r2} = 1$, $\varepsilon_{r1} = 1$, $\varepsilon_{r2} = 10{-}j20$. **d** $\mu_{r1} = 1$, $\mu_{r2} = 10{-}j5$, $\varepsilon_{r1} = 1$, $\varepsilon_{r2} = 2$. *After* [2]

where Z_{Fi} is the wave impedance, given by,

$$Z_{Fi} = \sqrt{\frac{\mu_0 \mu_{ri}}{\varepsilon_0 \varepsilon_{ri}}} \tag{4.19}$$

A few important reflection conditions are known by specific names. For example, the condition in which the angle of incidence and the angle of reflection are equal, $\theta_r = \theta_i$, is called *Snell's law of reflection*, whereas the condition $\sqrt{\varepsilon_{r1}\mu_{r1}}\,sin\theta_i = \sqrt{\varepsilon_{r2}\mu_{r2}}\,sin\theta_t$, is called *Snell's law of refraction*. In addition, the condition,

$$R_{\parallel} = 0 \tag{4.20}$$

when there is no reflection, is set by the angle given by,

$$sin\,\theta_{iB} = \sqrt{\frac{\left(\dfrac{\varepsilon_{r2}}{\varepsilon_{r1}} - \dfrac{\mu_{r2}}{\mu_{r1}}\right)}{\left(\dfrac{\varepsilon_{r2}}{\varepsilon_{r1}} - \dfrac{\varepsilon_{r1}}{\varepsilon_{r2}}\right)}} \tag{4.21}$$

called *Brewster angle*. For $\varepsilon_{r1} = \varepsilon_{r2} = real$ and $\mu_{r1} = \mu_{r2} = 1$, the Brewster angle is given by,

$$\theta_{iB} = arctan\left(\frac{\varepsilon_{r2}}{\varepsilon_{r1}}\right) \tag{4.22}$$

In the TE case, Fig. 4.8b, the Fresnel reflection coefficients take the form,

$$\vec{E}' = R_\perp \vec{E}^i, \; R_\perp = \frac{Z_{F2}\,cos\theta_i - Z_{F1}\,cos\theta_t}{Z_{F2}\,cos\theta_i + Z_{F1}\,cos\theta_t} \tag{4.23}$$

$$\vec{E}' = T_\perp \vec{E}^i, \; T_\perp = \frac{2Z_{F2}\,cos\theta_i}{Z_{F2}\,cos\theta_i + Z_{F1}\,cos\theta_t} \tag{4.24}$$

where Z_{Fi} is the wave impedance, given by,

$$Z_{Fi} = \sqrt{\frac{\mu_0\mu_{ri}}{\varepsilon_0\varepsilon_{ri}}} \tag{4.25}$$

Snell's law of reflection remains, $\theta_r = \theta_i$, but the Brewster angle condition,

$$R_\perp = 0 \tag{4.26}$$

gives the Brewster angle,

$$sin\,\theta_{iB} = \sqrt{\frac{\left(\dfrac{\mu_{r2}}{\mu_{r1}} - \dfrac{\varepsilon_{r2}}{\varepsilon_{r1}}\right)}{\left(\dfrac{\mu_{r2}}{\mu_{r1}} - \dfrac{\mu_{r1}}{\mu_{r2}}\right)}} \tag{4.27}$$

which, at the condition $\varepsilon_{r1} = \varepsilon_{r2} = 1$ and $\mu_{r1} = \mu_{r2} = real$, results in,

$$\theta_{iB} = arctan\left(\frac{\mu_{r2}}{\mu_{r1}}\right) \tag{4.28}$$

Calculation of the typical behavior of these reflection and transmission coefficients are shown in Fig. 4.9.

Many other reflection scenarios are possible, for example, multilayer situations, in which medium 2 is a stack of materials with alternating dielectric properties. These are usually dealt with using numerical schemes, and will not be discussed here [2].

Another model of interest is the two-ray ground reflection model, Fig. 4.10a. Figure 4.10b, c shows the results of calculations of the path loss for propagation from transmitter to receiver for vertically and horizontally polarized antennas, respectively [2].

A calculation of the path loss incurred as a signal propagates from transmitter to receiver produces the results in Fig. 4.11 [2].

4.1.3.3   Diffraction Models

The phenomenon of *diffraction* is responsible for the propensity of radio signals to propagate behind obstructions, and around the curved surface of the earth. The modeling of this phenomenon is typified by Fig. 4.12a b, which characterize the wave as rays which, when blocked by an object, results in a shadow region, and Fig. 4.12c, d, in which there is no shadow.

Calculating the impact of diffraction on wave propagation is performed by a number of techniques. These include [2, 4–8]:

1. Empirical solutions;
2. Knife edge diffraction[1] (absorbing or metallic half plane, Huygens);
3. Approximations for multiple knife edges[1] [5–8];
4. High frequency methods
   - GTD: Geometrical Theory of Diffraction
   - UTD: Uniform Geometrical Theory of Diffraction
   - PTD: Physical Theory of Diffraction
   - ITD: Incremental Theory of Diffraction

5. Numerical solutions
   - PEM: Parabolic Equation Method
   - IEM: Integral Equation Method

---

[1] no separation of diffracted field

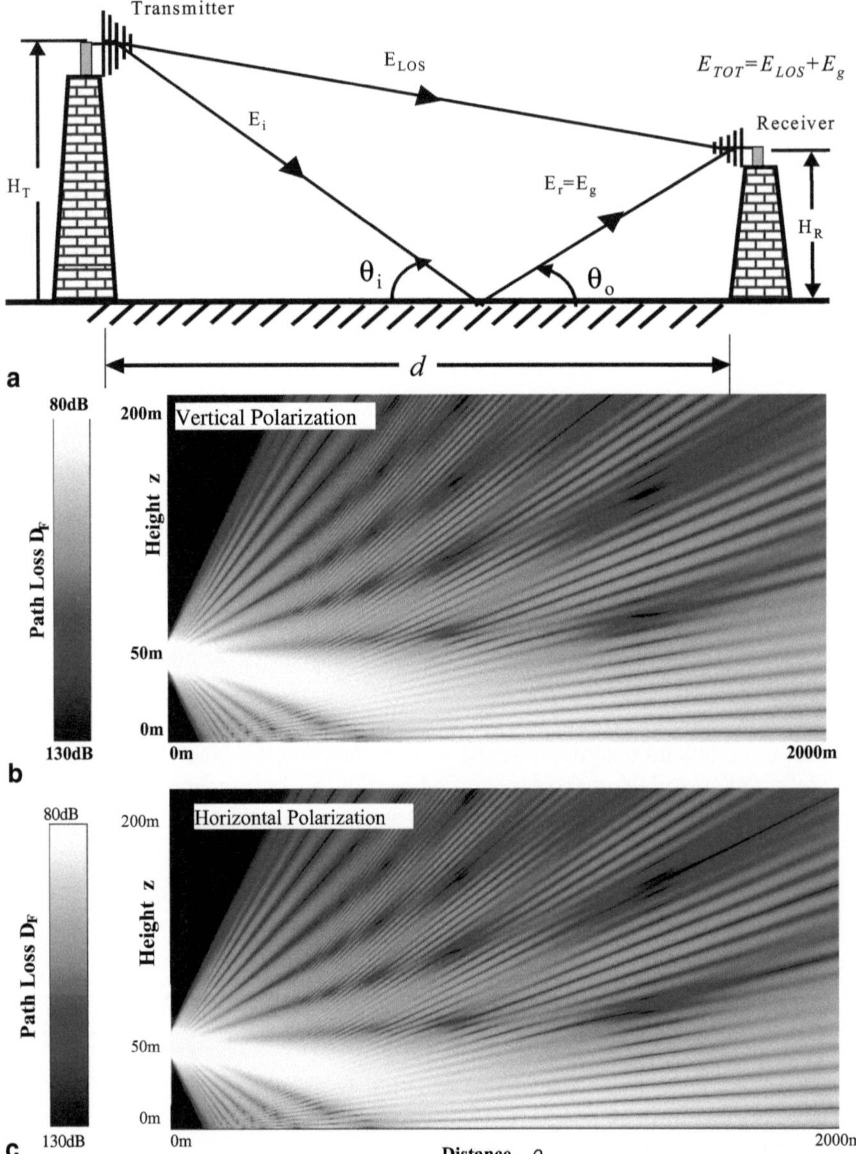

Fig. 4.10 a Two-ray ground reflection model. Path loss of transmit antenna through an ideal conductive metallic half-space for frequency of 500 MHz (transmit antenna, $z_T = 50$ m). **b** Vertically polarized antenna. **c** Horizontally polarized antenna. *After* [2]

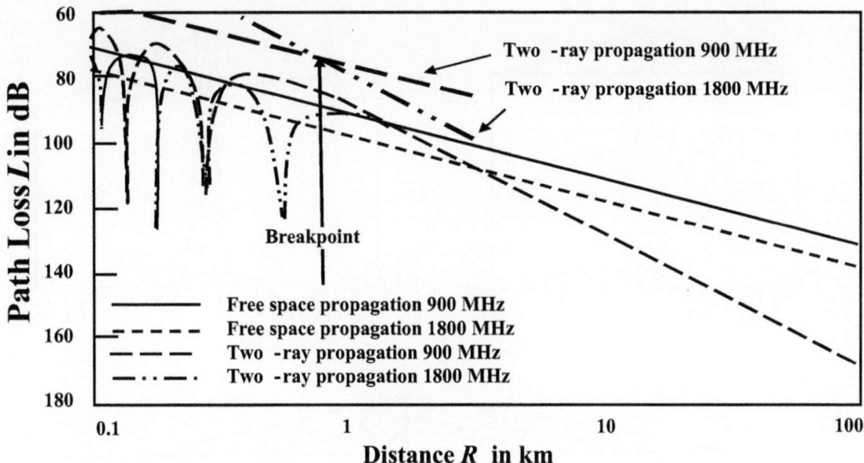

Fig. 4.11 Two-path propagation path loss L for GSM, for propagation in horizontal polarization mode. The geometrical parameters are $H_T = 30$ m; $H_R = 1.5$ m. The path loss exponent changes from 2 to 4 at the breakpoint. *After* [2]

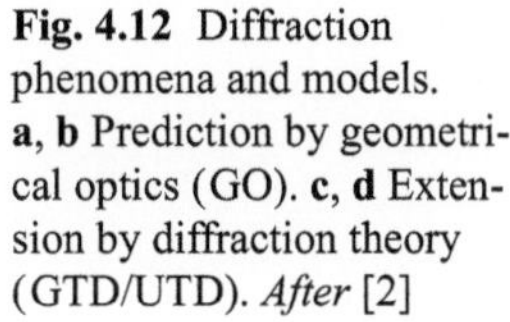

Fig. 4.12 Diffraction phenomena and models. **a, b** Prediction by geometrical optics (GO). **c, d** Extension by diffraction theory (GTD/UTD). *After* [2]

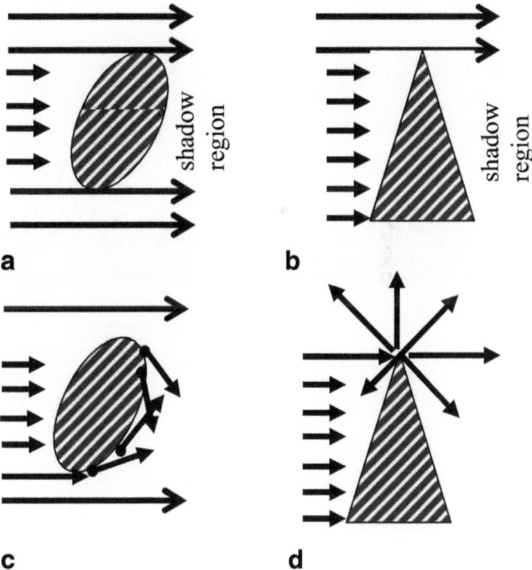

Two models are used to capture the knife's edge geometry, whose diffraction properties are abstracted to concentric circles for defining the boundaries of successive Fresnel zones, see Fig. 4.13.

Various diffraction scenarios can be manifested, as depicted in Fig. 4.14.

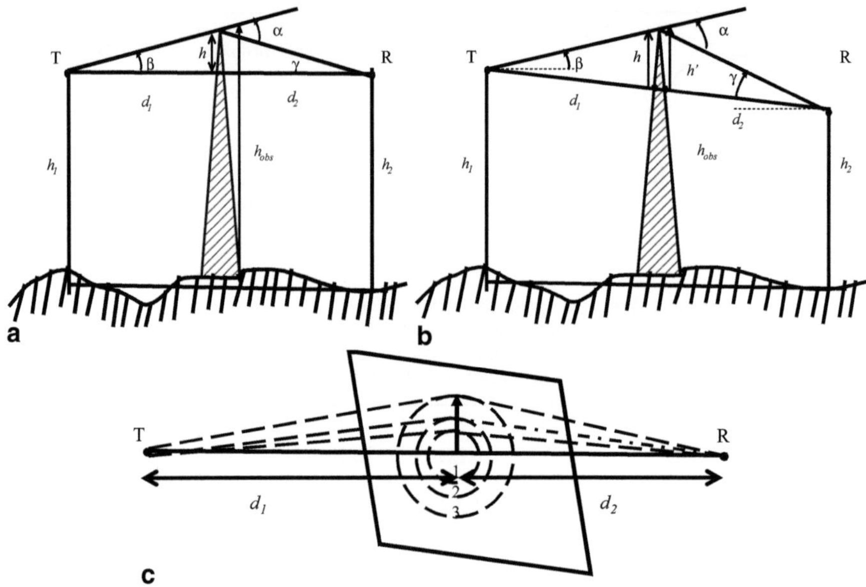

Fig. 4.13 Knife-edge diffraction geometry. **a** Infinite obstruction blocking line-of-sight path with similar T, R heights. **b** Unequal T, R heights. **c** Concentric circles which define boundaries of successive Fresnel zones. (The distances between the diffracted path difference between the zones is $n\lambda/2$). *After* [3]

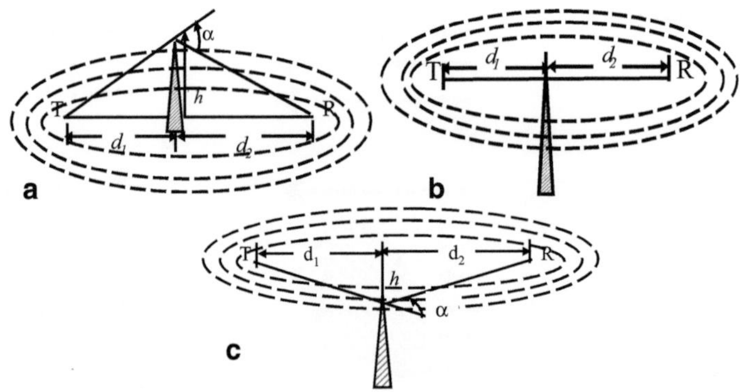

Fig. 4.14 Illustrations of Fresnel zones for various knife-edge diffraction scenarios. *After* [3]

Mathematically, the field strength of the knife-edge diffracted wave (E_d), relative to the free space field strength in absence of both ground and knife-edge (E_0) is given by [3],

$$\frac{E_d}{E_0} = F(\nu) = \frac{(1+j)}{2}\int_{\nu}^{\infty}\exp^{\frac{-j\pi t^2}{2}}dt \tag{4.29}$$

Table 4.1 Diffraction gain for various diffraction parameter values

Fresnel–Kirchoff diffraction parameter, v	G_d (dB) Approximation
$v \leq -1$	0
$-1 \leq v \leq 0$	$20\log(0.5 - 0.62v)$
$0 \leq v \leq 1$	$20\log(0.5\exp(-0.95v))$
$1 \leq v \leq 2.4$	$20\log(0.4 - \sqrt{0.1184 - (0.38 - 0.1v)^3})$
$v > 2.4$	$20\log\left(\dfrac{0.225}{v}\right)$

where, $F(n)$ is the complex Fresnel integral, and,

$$v = h\sqrt{\frac{2(d_1 + d_2)}{\lambda d_1 d_2}} \tag{4.30}$$

is the Fresnel–Kirchoff diffraction parameter. The impact of the knife's edge diffraction is captured by the so-called diffraction "Gain," G_d, given by,

$$G_d(dB) = 20\log|F(v)| \tag{4.31}$$

Table 4.1 gives various approximations to G_d as function to the diffraction parameter.

A calculation of the relative field strength of the absorbing knife's edge model is given in Fig. 4.15.

The calculated field strength, normalized to free space level, assuming an isotropic transmitting antenna and a semi-infinite absorbing knife edge is depicted in Fig. 4.16.

4.1.3.4   Scattering Models

There are three basic types of scattering, namely, point scattering, distributed scattering, which includes, rough surface scattering, and volume scattering, Fig. 4.17.

Point scattering characterizes simple targets, such as those with spherical and cylindrical shapes; these may, in general, be analyzed mathematically by exact methods [2]. Other shapes in this category, such as needle, spheroidal, and plates may be analyzed by low- and high-frequency approximations [2, 4–8].

Rough surface scattering is elicited by a surface height whose height/thickness is virtually random. In this case, analysis is done via numerical methods, or a mean value approximation that makes use of statistical information of the height while using the Kirchhoff and a scalar approximation, the Kirchhoff and a statistical phase approximation, the small perturbation method, or the phase perturbation method [2].

Volume scattering is caused by a relatively large and dense array of objects, such as the trees in a forest. Its analysis utilizes numerical methods that take into consideration the location and orientation of all scatterers, or a mean value approximation

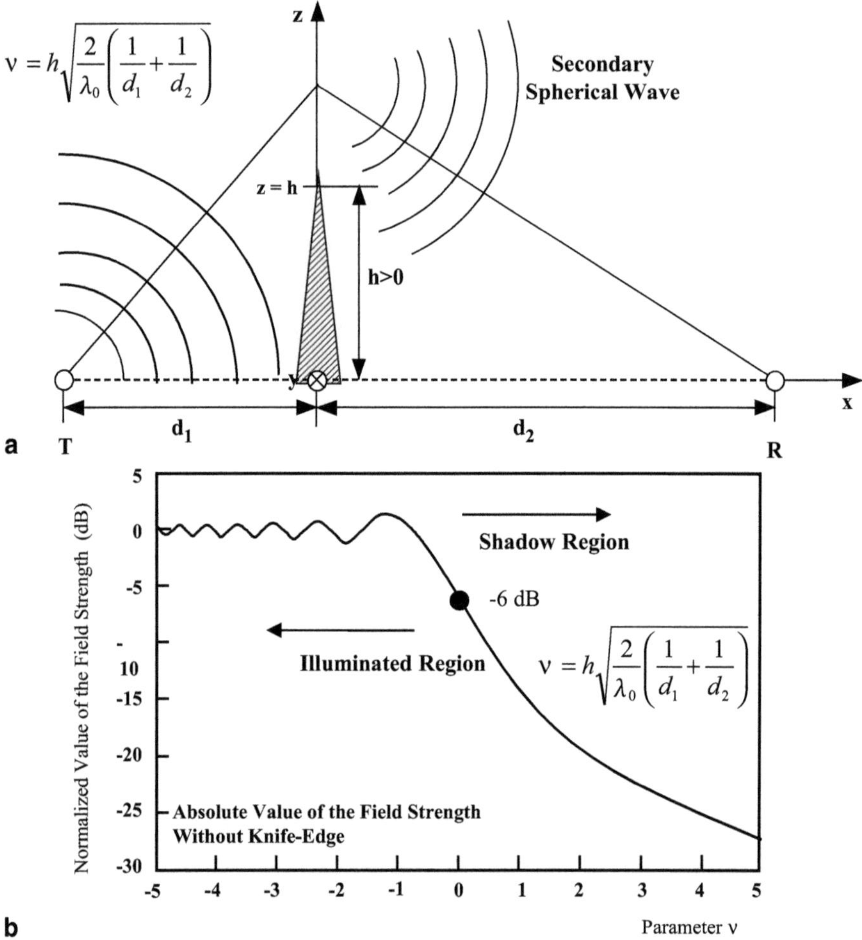

Fig. 4.15 Relative field strength |F/F0| at knife-edge-diffraction as a function of the parameter v.
After [2]

that makes use of statistical information and the Born approximation, the first-order renormalization, and radiative transfer theory [2, 4–8].

In terms of the far field scattering description, Fig. 4.18, the back scattering radar cross section of a sphere, considered as a point scatterer, is shown in Fig. 4.19 [2].

For vegetation, the volume scattering may be depicted as in Fig. 4.20.

The attenuation for a wave that propagates through vegetation of thickness d is modeled by,

$$\frac{D}{dB} = \eta \cdot \left(\frac{f}{1MHz}\right)^{v} \cdot \left(\frac{d}{1m}\right)^{\gamma} \tag{4.32}$$

where, η, v, and γ are functions of the vegetation and polarization. For example, a wave propagating through a forest with η=0.187, v=0.284, and γ=0.588 (which

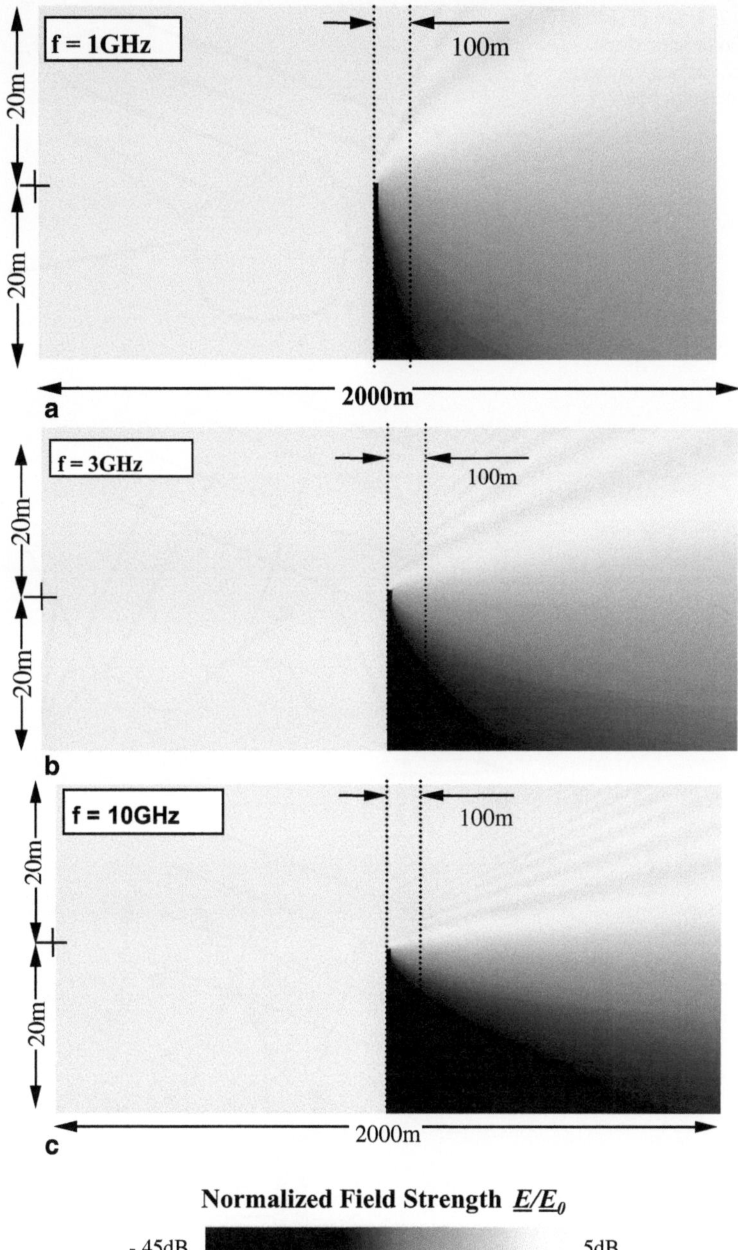

Fig. 4.16 Frequency dependence of single knife edge diffraction. **a** At 1 GHz. **b** At 3 GHz. **c** At 10 GHz. *After* [2, 4]

Fig. 4.17 Types of scattering. **a** Point scattering. **b** Rough surface scattering. **c** Volume scattering

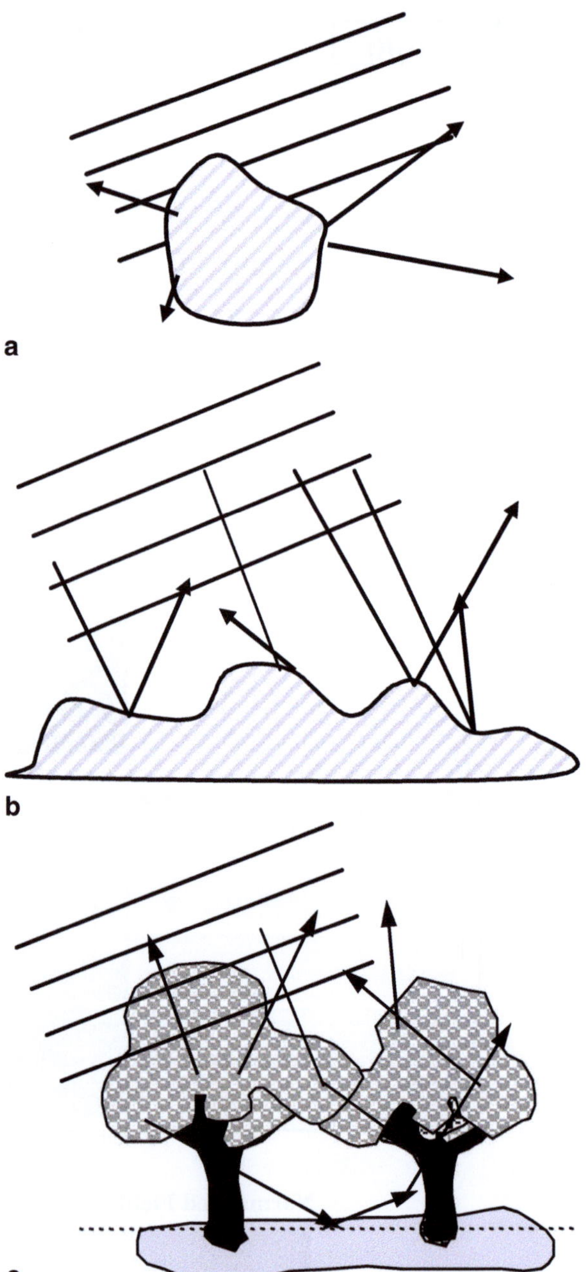

General scattering

$$\vec{E}^{Total} = \vec{E}^i + \vec{E}^s \quad \text{i=incident (without scatterer)}$$
$$\vec{H}^{Total} = \vec{H}^i + \vec{H}^s \quad \text{s=scattered=Total - incident}$$

Polarimetric far field scattering matric

$$\begin{pmatrix} E^s_\vartheta \\ E^s_\psi \end{pmatrix} = \frac{e^{-jkr}}{r} \begin{pmatrix} \underline{S}_{\vartheta\vartheta} & \underline{S}_{\vartheta\psi} \\ \underline{S}_{\psi\vartheta} & \underline{S}_{\psi\psi} \end{pmatrix} \begin{pmatrix} E^i_\vartheta \\ E^i_\psi \end{pmatrix}$$

Far field scattering description

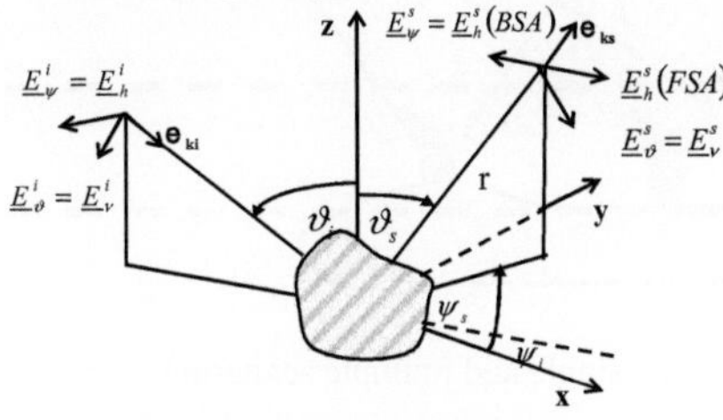

FSA: Forward scattering alignment convention
BSA: Backward scattering alignment convention

Polarimetric Radar cross section (RCS)

$$\sigma_{pq} = 4\pi r^2 \frac{\left|E^s_p\right|^2}{\left|E^t_q\right|^2} = 4\pi \left|\underline{S}_{pq}\right|^2 , \quad p,q = \vartheta,\psi$$

$$\sigma_{pq}\big/_{dBsm} = 10 \, \log \left(\sigma_{pq}\big/_{1m^2} \right)$$

$$S^s_p = \frac{\sigma_{pq} S^i_q}{4\pi r^2}, \quad S = power \;\; density \left[W\big/_{m^2}\right]$$

Normalized polarimetric scattering coefficients

$$\sigma^0_{pq} = \sigma_{pq}\big/_A = 4\pi\left|\underline{S}_{pq}\right|^2 \big/_A , \quad p,q = \vartheta,\psi$$

$$\sigma^{0,vol}_{pq} = \sigma_{pq}\big/_V = 4\pi\left|\underline{S}_{pq}\right|^2 \big/_V , \quad p,q = \vartheta,\psi$$

Fig. 4.18 Description of scattering mechanisms in terms of far field and RCS matrix for point scatterer

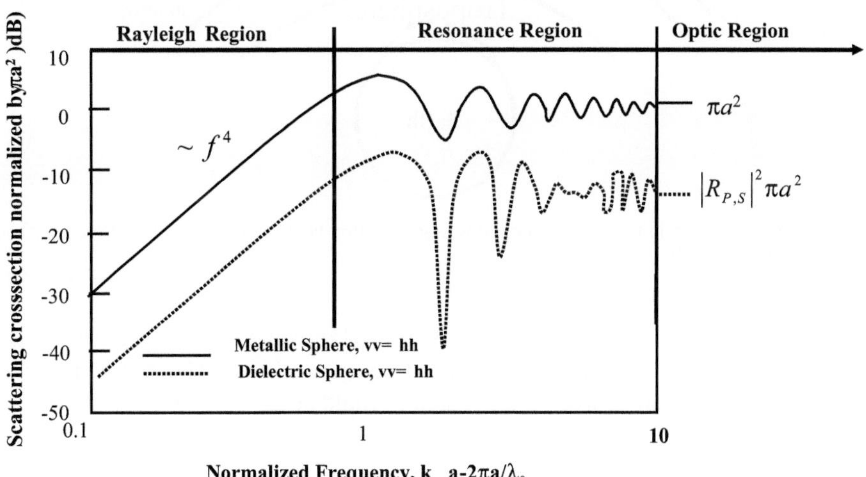

Fig. 4.19 Normalized back scattering radar cross-section $\sigma_{pq}/\pi a^2$ of a metallic and a dielectric sphere with relative permittivity $\varepsilon = 2 - j0.5$ as a function of normalized frequency $k_o a = 2\pi a/\lambda_0$ (angle of incidence and angle of reflection: $\theta_i = 90_o, \psi_i = 0^o, \theta_s = +/-180°$; cross-polarization: $\sigma_{vh} = \sigma_{hv} = 0$). *After* [2]

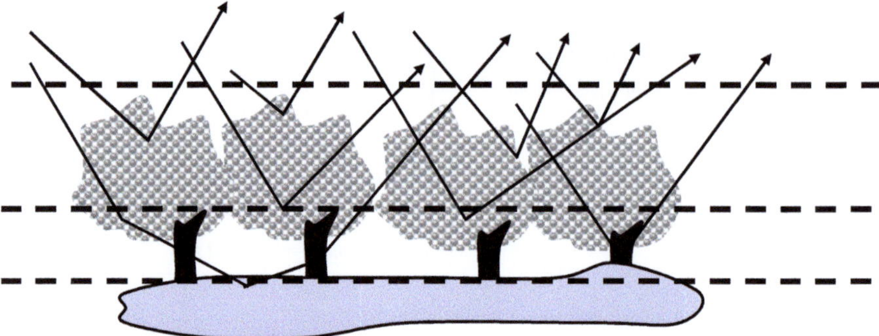

Fig. 4.20 Volume scattering in vegetation involves both single and multiple scattering

Atmosphere of the Earth

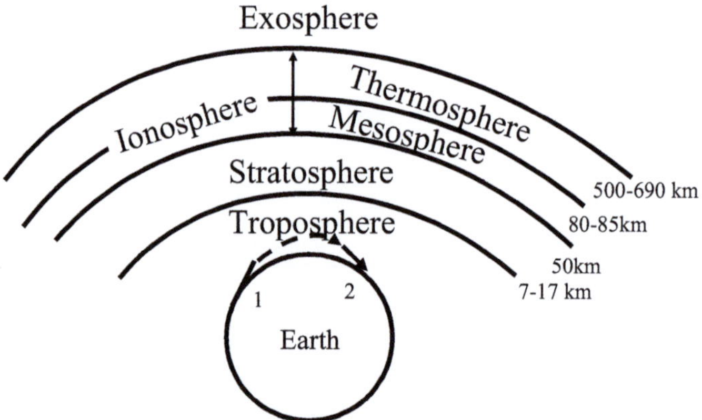

Fig. 4.21 Refraction in the troposphere allows propagation between points on the curved surface of the earth. *After* [2]

are valid for d<400 m, and 200 MHz<f<95 GHz) would result in an average attenuation of D=11.3 dB, for d=40 m at a frequency of 900 MHz. At frequency of 1800 MHz, the resulting average attenuation would be D=13.8 dB [2].

4.1.3.5   Refraction in the Troposphere

The earth's atmosphere is stratified in various layers, namely, troposphere, stratosphere, mesosphere, thermosphere, and exosphere. The varying index of refraction in the troposphere acts to refract/curve waves launched from the surface of the earth, Fig. 4.21, allowing communication that is NOT in line-of-sight with the transmitter.

4.2   Description of Antennas and Their Parameters

Antennas are transducers that couple the power output by the transmitter to, or the power received by the receiver from, space. In particular, at the transmitter, the last power amplifier is connected to a waveguide that is impedance-matched to an antenna which, in turn, realizes a transition region whose output is impedance-matched to free space, Fig. 4.22a, thus launching the transmitted electromagnetic wave. At a receiver, the opposite occurs. The antenna, which is impedance-matched to free space, provides a transition region to a waveguide whose output is impedance-matched to the receiver, Fig. 4.22b.

The frequency range of antennas utilized in wireless communications spans from about 10 kHz (wavelength $\lambda_0 = 30$ km) up to about 300 GHz (wavelength $\lambda_0 = 1$ mm). A variety of typical antennas is shown in Fig. 4.23 [2, 4]. In general, the type of antenna chosen depends on various design factors, in particular, the specific application, and the desired radiation properties. The weight, volume and mechanical stability also may play a role. The role of the antenna dimensions become more important as the wavelength decreases [2].

The fundamental performance parameters of an antenna are its directivity and gain, based on which its effective area and efficiency are defined [2]. These may be expressed analytically with respect to the spherical coordinate system in Fig. 4.24 [9].

In Fig. 4.24, P_n is the normalized antenna power pattern, defined as [9],

$$P_n(\theta,\phi) = \frac{P(\theta,\phi)}{P(\theta,\phi)_{\max}} \quad \text{(dimensionless)} \tag{4.33}$$

where, regardless of the wave polarization of the radiation, $P(\theta, \phi)$ is the radiated power given by,

$$P(\theta,\phi) = S_r r^2 = \frac{1}{2}\frac{E^2(\theta,\phi)}{Z}r^2 = \frac{1}{2}H^2(\theta,\phi)Zr^2 \quad (\text{Wsr}^{-1}) \tag{4.34}$$

with, S_r being the radial component of the Poynting vector,

$$S_{av} = \frac{1}{2}\int_s \mathrm{Re}(\vec{E}\times\vec{H}^*)\cdot d\vec{s} \tag{4.35}$$

$E(\theta, \phi)$ is the total transverse electric field as a function of angle, $H(\theta, \phi)$ is the total transverse magnetic field as a function of angle, r is the distance from the antenna to the point of measurement, and Z is the intrinsic impedance of the medium. Since P_n is dimensionless, integrating it with respect to angle (θ, ϕ) captures the solid angle occupied by the total radiated pattern solid angle or *beam solid angle*, Ω_A, which is given by [9],

$$\Omega_A = \int_{\phi=0}^{\phi=2\pi}\int_{\theta=0}^{\theta=\pi} P_n(\theta,\phi)\sin\theta\, d\theta\, d\phi = \iint_{4\pi} P_n(\theta,\phi)\, d\Omega \tag{4.36}$$

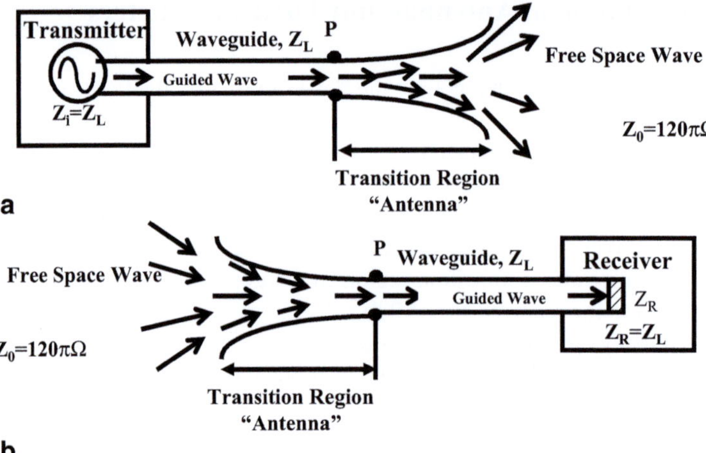

Fig. 4.22 Antenna as a transition region between guided wave and free-space wave. **a** Transmitter antenna. **b** Receiver antenna. *After* [2]

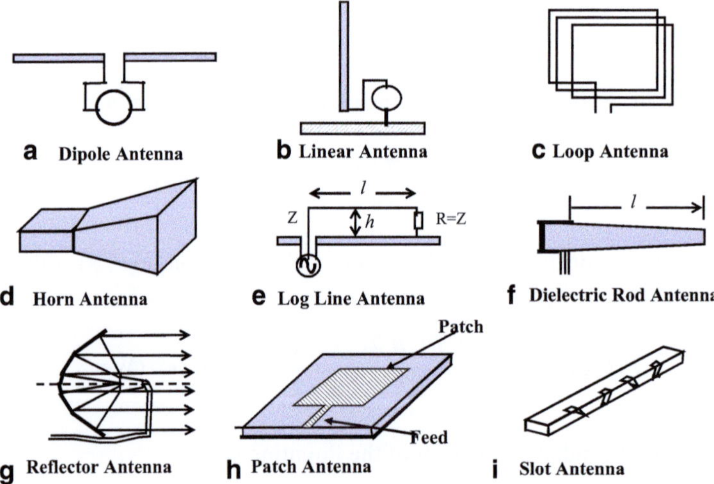

Fig. 4.23 Typical antennas. *After* [2]

The beam solid angle is interpreted as the *effective solid angle* through which all the power from a transmitting antenna would flow if this power were a constant equal to the maximum power throughout this angle [9]. This angle is usually approximately equal to the half-power beam width (HPBW), i.e., with respect to the maximum power at the center, this is the angle spanning either side of the maximum, at which the power has dropped to one-half the maximum power.

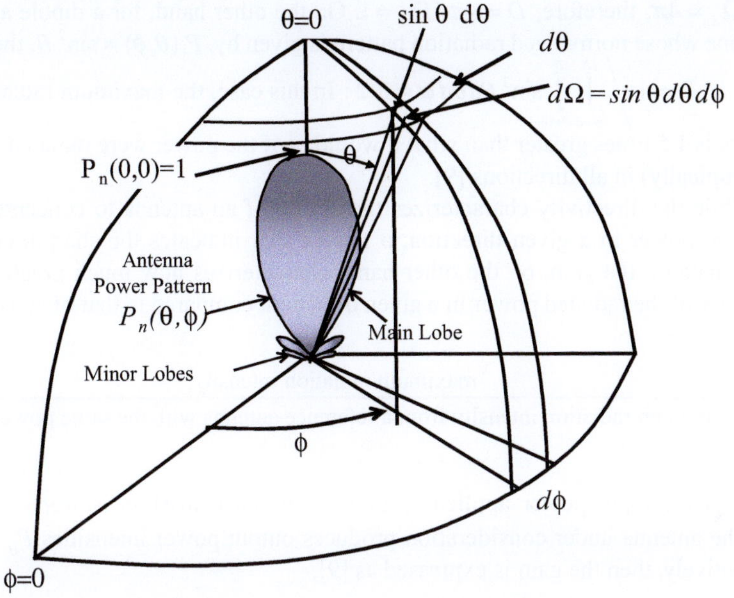

Fig. 4.24 Normalized antenna power pattern P_n with maximum aligned with the θ=0 direction (zenith)

The directivity, D, captures how well an antenna concentrates the radiated energy in a given direction, as opposed to other directions [10]. It is given by,

$$D = \frac{\text{maximum radiation intensity}}{\text{average radiation intensity}} = \frac{P(\theta,\phi)_{max}}{P_{av}} \quad \text{(dimensionless)} \quad (4.37)$$

The average radiation intensity is given by the total power radiated W divided by the maximum solid angle 4π sr. Therefore,

$$D = \frac{P(\theta,\phi)_{max}}{W/4\pi} = \frac{4\pi P(\theta,\phi)_{max}}{\iint\limits_{4\pi} P(\theta,\phi)d\Omega} = \frac{4\pi P(\theta,\phi)_{max}}{\iint\limits_{4\pi}[P(\theta,\phi)/P(\theta,\phi)_{max}]d\Omega} = \frac{4\pi}{\iint\limits_{4\pi} P_n(\theta,\phi)d\Omega}$$

$$(4.38)$$

Using (4.36), this may be expressed as,

$$D = \frac{4\pi}{\Omega_A} \quad \text{(dimensionless)} \quad (4.39)$$

Since D is inversely proportional to the beam solid angle Ω_A, it is concluded that, the smaller Ω_A, the greater the directivity, i.e., the greater is the ability of the antenna to concentrate the power into a narrow solid angle [9]. For example, for an isotropic antenna, i.e., one that radiates the same intensity in all directions, $P_n(\theta,\phi) = 1$

and $\Omega_A = 4\pi$, therefore, $D = 4\pi / \Omega_A = 1$. On the other hand, for a dipole antenna, i.e., one whose normalized radiation pattern is given by, $P_n(\theta, \phi) = \sin^2 \theta$, the *directivity is* $D = 4\pi \Big/ \displaystyle\int_{\psi=0}^{2\pi} \int_{\theta=0}^{\pi} \sin^2 \theta d\theta d\psi = 3/2$. In this case, the maximum radiation intensity is 1.5 times greater than what it would be if the power were radiated equally (isotropically) in all directions [9].

While the directivity characterizes the ability of an antenna to concentrate the radiated power in a given direction, a feature that indicates the sharpness of the beam pattern, the gain, on the other hand, characterizes how much greater is the intensity of the radiated power in a given direction, compared to that of an isotropic antenna, i.e.,

$$G = \frac{\text{maximum radiation intensity}}{\text{maximum radiation intensity from a reference antenna with the same power input}}$$

$$(4.40)$$

If the same input power applied to both a reference lossless isotropic antenna and the antenna under consideration produces output power intensities P_0 and P', respectively, then the gain is expressed as [9],

$$G_i = \frac{P'}{P_0} \tag{4.41}$$

The actual radiated power P' is related to that of a 100 % efficient antenna, P_{max}, by the efficiency, η, as follows,

$$P' = \eta P_{max} \tag{4.42}$$

where $\eta < 1$. Substituting (4.42) into (4.40), one obtains,

$$G_i = \eta \frac{P_{max}}{P_0} = \eta \cdot D \tag{4.43}$$

Therefore, if the efficiency of an antenna is 100 %, then its gain and directivity coincide. When defined with respect to an isotropic source, it is customary to express the gain and directivities in *decibels over isotropic*, or *dBi*, where,

$$G_i(dBi) = 10 \log G_i \tag{4.44}$$

and similarly for the directivity [9].

The concept of *antenna effective area* arises when we consider a receiving antenna, i.e., one that *collects* energy. In this case, it is thought that the receiving antenna possesses an aperture of area A, which captures energy from a passing electromagnetic wave. This aperture area is defined as,

$$A = \frac{\text{received power}}{\text{power density of the incident wave}} = \frac{I^2 R_L}{S} \tag{4.45}$$

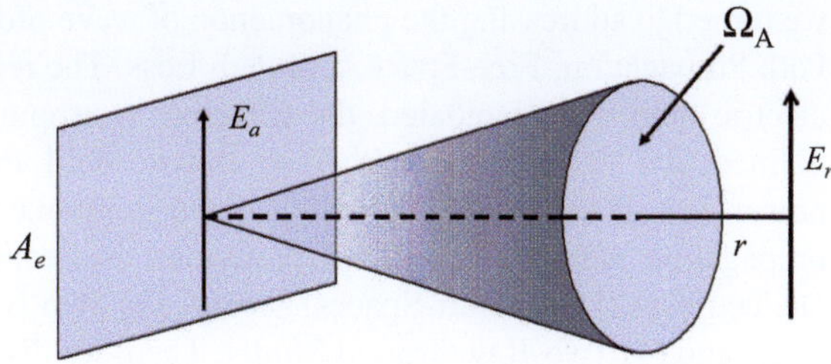

Fig. 4.25 Radiation from aperture A_e with uniform field E_a

where A is the aperture (m²), I is the rms terminal current (A),

$$I = \frac{V}{Z_L + Z_A} \tag{4.46}$$

where, V is the rms emf induced in the antenna by the passing wave, and Z_A and Z_L are the antenna and load impedances, respectively, S is the Poynting vector of the incident wave (W/m^2), and R_L is the load resistance (Ω). It can be shown [9] that the condition for the maximum aperture occurs when the antenna is oriented for maximum response emf V, it is matched ($Im\{Z_L\} = -Im\{Z_A\}$), and it is lossless ($Re\{Z_A\} = R_r$), where R_r is the antenna radiation resistance. In that case the aperture area A is called effective antenna area A_e, and is given by,

$$A_e = \frac{V^2}{4SR_r} \tag{4.47}$$

Now, it can be shown that an antenna with effective aperture A_e has a beam solid angle corresponding to Ω_A, Fig. 4.25, both of which are related to the wavelength by [9],

$$\lambda^2 = A_e\Omega_A \ (m^2) \tag{4.48}$$

From this relationship, substitution of it in the expression for directivity gives,

$$D = \frac{4\pi}{\lambda^2} A_e \tag{4.49}$$

So, the directivity is inversely proportional to the wavelength.

4.3 Summary

In this chapter, we have presented a discussion of the fundamentals of the radio channel. We began by reviewing typical radio transmitter and receiver block diagrams, together with the components of a wireless communications system. Then, we turned to addressing the phenomenon of wave propagation, in particular, Multi-Path Propagation, Free-Space, and Path Loss. The relationship between power and electric field was elucidated, the Effective Isotropic Radiated Power (EIRP) was defined, and the measured received electric field and its relation to the received power was given. Then, the various models invoked in calculating the effects of propagation were presented. In particular: Reflection at: (1) Two Half-Spaces: Dielectrics; (2) Two Half-Spaces: Metals; (3) Two Half-Space: Multi-Layer Structures; and (4) Two-Ray Ground Model. Then, we discussed Diffraction, in particular, from a Knife-Edge. We concluded this discussion with the topics of Scattering, and Reflection in the Troposphere. The chapter was concluded with a discussion of antenna parameters, in particular: (1) Efficiency; (2) Effective Area; (3) Gain; and (4) Directivity.

References

1. H. L. Krauss, C. W. Bostian, and F. H. Raab, *Solid State Radio Engineering*, John Wiley & Sons, Inc., 1980.
2. Prof. Dr.-Ing. W. Wiesbeck, Skriptum zur Vorlesung, *Grundlagen der Hochfrequenztechnik*, Universität Karlsruhe (TH), Institut für Höchstfrequenztechnik und Elektronik, 2007.
3. T. S. Rappaport, *Wireless Communications: Principles and Practice*, Second Ed., Prentice-Hall, Inc. 2002
4. Dipl.-Ing. Thomas Fügen, Lecture Notes, Course, "Wave Propagation and Radio Channels for Mobile Communications," KIT, IHE, SS2010.
5. Vogler L. E. "An attenuation function for multiple knife-edge diffraction", Radio Science, vol. 17, pp. 1541–1546, 1982.
6. Epstein, j and D. W. Peterson, "An experimental study of wave propagation at 850 Mc," *Proc. IRE*, vol. 41, No. 5, 1953, pp. 595–611.
7. C. Huygens, *Traité de la Lumiere* (completed in 1678, published in Leyden in 1690).
8. K. Furutsu, "A systematic theory of wave propagation over irregular terrain", *Radio Science*, vol. 17, No. 5, pp. 1037–1059, May, 1982.
9. J. D. Kraus and K. R. Carver, *Electromagnetics*, Second Edition, McGraw-Hill, Inc., New York, 1973.
10. W. L., Stutzman and G. A. Thiele, *Antenna Theory and Design*, Second Edition, John Wiley & Sons, Inc., New York (1998).

Chapter 5
Noise, Nonlinearity and Time Variance

Abstract In this chapter we deal with the topics of Noise, Nonlinearity and Time Variance. We begin with a presentation of the various noise sources of importance to a receiver, in particular, Thermal Noise, and its definition, quantification and voltage and current circuit models; and Shot noise and its quantification. Then, we take on the topic of Signal-to-Noise Ratio, including, available noise power, Noise Figure, Noise Temperature, and Noise Figure of Cascaded Linear Networks. This is followed by the topic of Mixer Noise Figure, in particular, Single-Sideband and Double-Sideband. We conclude this part with a discussion of the calculation of the noise at baseband for a system consisting of a cascade of building blocks. Then, we address the topics of Nonlinearity and Time Variance. In particular, we review effects of nonlinearity, namely, Gain Compression and Intermodulation, and then address the topics of Cascaded nonlinear stages, Second-order intercept point, Intermodulation Distortion Formulas, Cascade Second-order input intercept point (IIP2), Twotone IIP2, One-tone IIP2, Time-varying-envelope signal IlP2, Thirdorder intercept point (IIP3), Cascaded third-order input intercept point (IIP3), and Two-tone IlP3.

5.1   Introduction

In this chapter, we deal with the topics of Noise, Nonlinearity and Time Variance. These are phenomena which, originating at the component level, play an important role in determining the overall system performance. In particular, by relating the overall system performance parameters to those characterizing the individual building blocks, the tradeoffs that optimize overall performance become transparent and aid in deriving the specifications of the individual building blocks.

5.2   Thermal Noise

In an electrical network we encounter conductors that contain a large number of free electrons and ions bound by molecular forces [1, 2]. The ions vibrate randomly about their normal (average) positions, however, this vibration is a function of

H. J. De Los Santos et al., *Radio Systems Engineering,*
DOI 10.1007/978-3-319-07326-2_5, © The Authors 2015

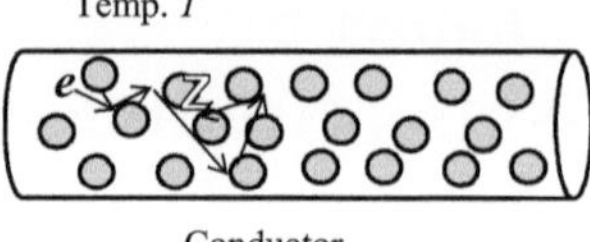

Fig. 5.1 Situation of randomly moving electrons colliding with vibrating ions in a conductor at temperature T

temperature. Collisions between the free electrons and the vibrating ions continuously take place and, as a result, there is a continuous transfer of energy between electrons and ions [1, 2], Fig. 5.1. This gives rise to the *resistance* of the conductor.

The freely moving electrons constitute a current which, over a long period of time, averages to zero, since many electrons on the average move in one direction as in another [1]. There are random fluctuations about this average, however, and, in fact, the *mean-squared fluctuations* in current are proportional to kT; $k = 1.38 \times 10^{-23}$ J/K is Boltzmann's constant [1]. The current fluctuations, in turn, capture the chaotic motion of particles possessing thermal energy. In the scenario discussed, there are no forces present inducing this motion in preferred directions, therefore, it is *spontaneous*.

5.2.1   *Quantification of Thermal Noise*

Thermal noise was first experimentally studied in 1928 by J. B. Johnson, and theoretically studied by H. Nyquist [2]. They demonstrated that a metallic resistor could be considered the source of spontaneous fluctuation voltages with mean-squared value:

$$\overline{v^2} = 4kTRB \tag{5.1}$$

where T is the temperature of the resistor in Kelvin, R is its resistance in ohms, and B is any arbitrary bandwidth [1]. This expression for the mean-squared thermal noise due to a resistor R implies that the noise is *white* (is independent if frequency). This may be shown to be correct up to high frequencies of the order of 10^{13} Hz [1]. The *thermal noise spectral density* is given by [1],

$$G_v(f) = \frac{\overline{v^2}}{2B} = 2kTR \tag{5.2}$$

This equation implies that resistors may be considered as sources of *noise voltage* [1, 2].

Two circuit noise models are employed in representing noise sources in circuits, namely, a voltage model and a current model, Fig. 5.2, where R is assumed to be noise-free, with the noise effect lumped into the noise-voltage, or current, source [1]. For example: Let B=4 kHz, T=293 K (200 °C), and R=10 kΩ. Then, $\overline{v^2} = 0.65 \times 10^{-12}$ *volts*^{2} or $\sqrt{\overline{v^2}} = 0.80$ μV rms. Notice that, for example, if the bandwidth is quadrupled to 16 kHz, the rms noise voltage and current are doubled to

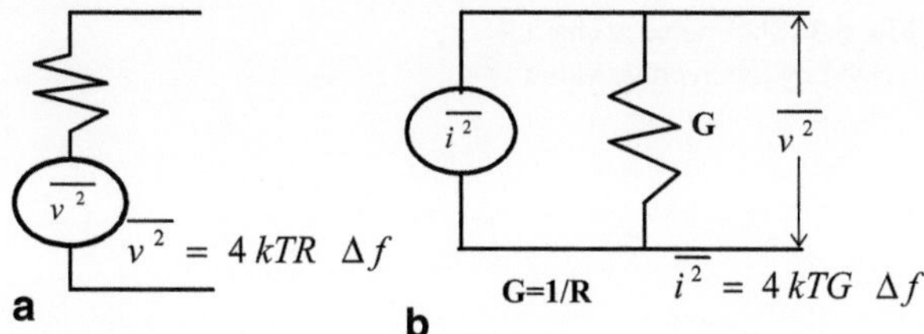

Fig. 5.2 Circuit models for representing noise. **a** Voltage model. **b** Current model

1.6 µV and 1.6×10^{-10} A, respectively. In these expressions, since R is typically temperature-dependent, care must be taken to utilize the resistance at the pertinent value of T. In summary: a resistance R at temperature T is found to be the source of a fluctuation (noise) voltage with mean-squared value $\overline{v^2} = 4kTRB$ and spectral density $G_v(f) = 2kTR$. For the current-noise model, we have, $\overline{i^2} = 4kTGB$ and spectral density, $G_v(f) = 2kTG$. B is the *noise-equivalent bandwidth* of the measuring instrument or circuit used to measure these quantities [1].

5.3   Shot Noise

Shot noise originates in fluctuations in the average carrier flow in some direction as a result of the discrete nature of matter [1]. The flow is typified by: (1) Electrons flowing between cathode and anode in a cathode-ray tube; (2) Electrons and holes flowing in semiconductors. The mechanism causing the fluctuation depends on the particular flow process, in particular, in a vacuum tube case it is the random emission of the electrons from the cathode, whereas in the semiconductor case it is the randomness in the number of electrons that continually recombine with holes, or on the number that diffuse, etc. [1]. It is found that the mean-squared fluctuations about the average value are proportional to the average value itself. Thus, shot noise is characterized by a dependence of the *noise* on the *average value*.

5.3.1   Quantification of Shot Noise

The shot noise formula is given by [1, 2],

$$\overline{i^2} = 2eI_{dc}\Delta f \tag{5.3}$$

Notice that the rms current is proportional to the square root of the current and BW (Fig. 5.3).

Widening the BW in circuit applications is counterproductive because it increases the amount of noise power available. For example: Let $I_{dc} = 1$ mA and $\Delta f = 4$ kHz. Then, $\overline{i^2} = 2(1.6 \times 10^{-19}) \times 10^{-3} \times (4 \times 10^3) = 1.3 \times 10^{-18}$ A^2, and the rms current generated is 1.1×10^{-9} A rms. If the current flows through a 5 kΩ resistor, the rms noise

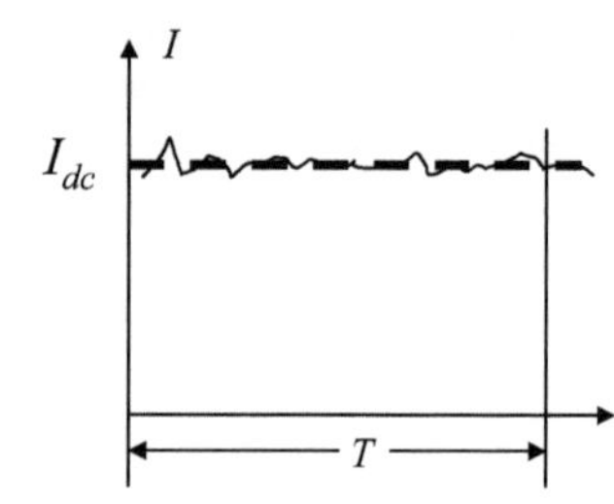

Fig. 5.3 Shot noise is characterized by the average value

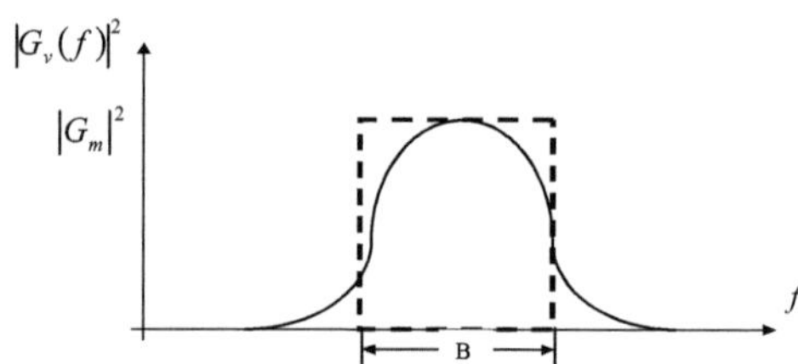

Fig. 5.4 Typical frequency response of amplifier with no sharp cutoff

voltage due to this current is 5.7 µV rms. If the BW quadrupled to 16 kHz, the rms noise voltage doubles to 11.3 µV rms.

The shot noise spectral density is given by,

$$G(f) = eI_{dc} \tag{5.4}$$

5.4   Noise Bandwidth

The most common sources of noise (thermal and shot noise) have a *uniform* spectral density. Therefore, the noise transmitted through an amplifier is determined by its bandwidth. If the amplifier frequency response were a rectangular box, it would be easy to determine its BW for noise calculations. However, the general amplifier frequency response does not have sharp cutoff, Fig. 5.4.

The *noise bandwidth*, then, is defined as the BW of the ideal frequency response "box" whose area is equal to that of the amplifier's frequency response, and is mathematically given by,

$$B = \frac{\int_0^\infty |G_v(f)|^2 \, df}{|G_m|^2} \tag{5.5}$$

Typically, B is approximately equal to the 3 dB bandwidth.

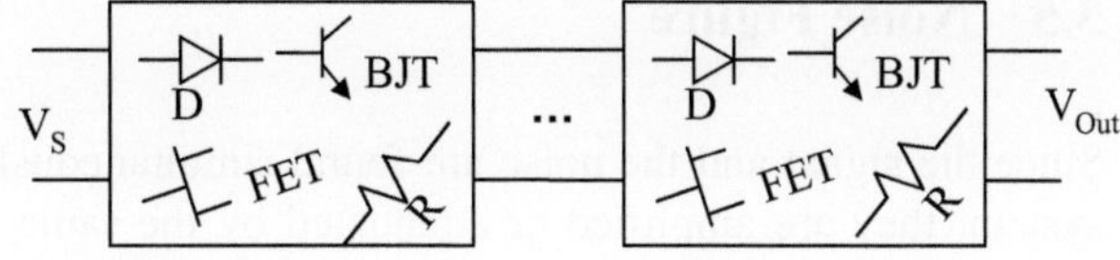

Fig. 5.5 Each stage of a system adds noise to the signal passing through

5.4.1   Signal-to-Noise Ratio

In a specified bandwidth, the signal-to-noise ratio (SNR) is defined as the ratio of the signal power to noise power at a port,

$$SNR = \frac{P_S}{P_N} = \frac{V_s^2}{V_n^2} \tag{5.6}$$

where V_s and V_n are the rms signal and noise voltages, respectively. In decibels,

$$SNR(dB) = 10 \log_{10} \frac{P_S}{P_N} \tag{5.7}$$

This expression is interpreted as follows: The larger the SNR, the less a signal is "corrupted" by the noise. The lowest permissible value of SNR depends upon the application: (1) $SNR_{min} = 10$ dB at the detector input of AM receiver; (2) $SNR_{min} = 12$ dB at the detector input of FM receiver; (3) $SNR_{min} = 40$ dB at the detector input of TV receiver [3].

As a signal passes through a cascade of amplifier stages, the SNR continually decreases because each stage adds noise, Fig. 5.5. In most systems, however, the amplified output noise is due primarily to: (1) Noise present along with the input signal; (2) Noise contributed by the first two stages (such as the RF amplifier and mixer stages in a receiver) [3].

5.4.2   Available Noise Power

The available power P_a of a source, is the maximum power that can be drawn from the source [3]. If the source has an internal impedance $Z_s = R + jX$, the maximum power will be delivered to a conjugate-matched load ($Z_L = R - jX$). If the open-circuit voltage of the source is V, the maximum power transfer theorem yields:

$$P_a = \frac{V^2}{4R} \tag{5.8}$$

Generally, it is the noise introduced at the input to a system that is most critical in determining the resultant SNR [3].

5.5   Noise Figure

Since the signal and the noise are found simultaneously at the input of a receiver system, they are amplified or attenuated by the same amount in passing through successive stages of the system. Ideally, therefore, we would like to deal with systems which introduce no noise in addition to that present at the input. This would imply a constant SNR throughout the system, since both signal and noise voltages are amplified or attenuated by the same amount. In practice, however, it is impossible to maintain a constant SNR, since any additional resistors or transistors, among other devices, encountered by the signal in its way from the input towards the output introduce additional noise. The SNR must, therefore, continuously decrease throughout the system [3]. However, the decrease may be negligible if the amplification at the front end of the system is made large enough so that the effect of the additional noise introduced in the subsequent stages is negligible. A study of SNR at different points in a system enables us to pinpoint the significant contributors to the noise and, therefore, it also enables us to design circuits and systems to minimize noise. It is, thus, important to provide a measure of the "noisiness" of a system, and this is accomplished by defining the *noise figure* parameter, F [3].

The noise figure F is defined as the ratio of the SNR S_s/N_s at the system input (source) to the SNR S_o/N_o at the system output [3]. It is given as,

$$NF = \frac{input\ SNR}{output\ SNR} = \frac{P_{si}/P_{ni}}{P_{so}/P_{no}} = \frac{P_{si}}{P_{so}} \cdot \frac{P_{no}}{P_{ni}} = \frac{1}{G_a} \cdot \frac{P_{no}}{P_{ni}} = \frac{G_a P_{ni} + P_{ne}}{G_a P_{ni}} = 1 + \frac{P_{ne}}{G_a P_{ni}} \qquad (5.9)$$

where G_a is the available power gain of the system, P_{si} and P_{ni} are the input signal and noise powers, respectively, P_{so} and P_{no} are the corresponding output signal and noise powers, and P_{ne} is the noise power introduced by the system that appears at the output. The value of NF is often expressed in decibels through the relation:

$$NF_{dB} = 10\log_{10} NF \qquad (5.10)$$

For a noise-free network, the input and output *SNRs* will be equal and *NF = 1 or NF_{dB} = 0.*

The *NF* may also be (equivalently) defined as the ratio of the actual noise power available from a noisy network to that which would be available if the network were noiseless [3].

$$NF = \frac{P_{no}}{G_a P_{ni}} \qquad (5.11)$$

5.5.1   *Spot Noise Figure*

The spot noise figure, as the name implies, is the noise figure at a selected frequency. It is defined as the ratio of: (1) The total available noise power per unit bandwidth at the output port to; (2) The portion thereof produced at the input frequency by the input termination, whose noise temperature is 290 K. If the network provides a conjugate match at the input port, the available power from the standard-temperature source in a 1-Hz bandwidth is equal to kT_0. The spot noise figure is given by [3]:

$$NF = \frac{P_{no}}{G_a(f)kT_0} \tag{5.12}$$

In practice, the value of P_{no} is measured over a BW Δf that is more than 1 Hz. The equation for NF then becomes:

$$NF = \frac{P_{no}}{G_a(f)kT_0\Delta f} \tag{5.13}$$

5.5.2   Noise Temperature

The advent of radio astronomy, and the impetus it gave to the development of low noise receivers, resulted in the development of amplifiers (masers and parametric amplifiers) with noise figure as low as 1.1 [1]. These devices require extremely low temperatures and special cooling systems for their effective operation. In these cases, the concept of noise figure no longer has a real significance, since it is the external noise introduced at the antenna that provides the major source of noise, rather than the internal receiver noise. This leads to the concept of *noise temperature.*

The noise temperature of any port of a network is defined as follows: A noise source that has an available power P_a in a small frequency interval Δf has an equivalent temperature equal to:

$$T_e = \frac{P_o}{k\Delta f} \tag{5.14}$$

In order to obtain a standard value for NF, the source temperature must be assumed to be $T_0 = 290$ K. For example, at $T = T_0 = 290$ K, $kT_0B = 4.00 \times 10^{-21}$ W/Hz ($= -204.0$ dBW/Hz $= -174.0$ dBm/Hz).

5.5.3   Effective Noise Temperature of a Network

If a thermal source of temperature T is connected to a noiseless network with small bandwidth Δf and available gain $G_a(f)$, the available noise power from the source is [1, 2],

$$P_{ns} = kT\Delta f \quad \text{(Watts)} \tag{5.15}$$

and the available noise power at the output of the network is,

$$P_{no} = G_a(f)kT\Delta f \tag{5.16}$$

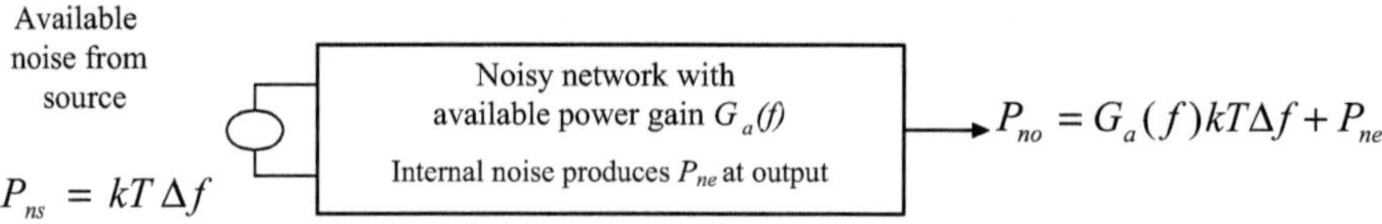

Fig. 5.6 Noisy network adds noise P_{ne} to input noise source

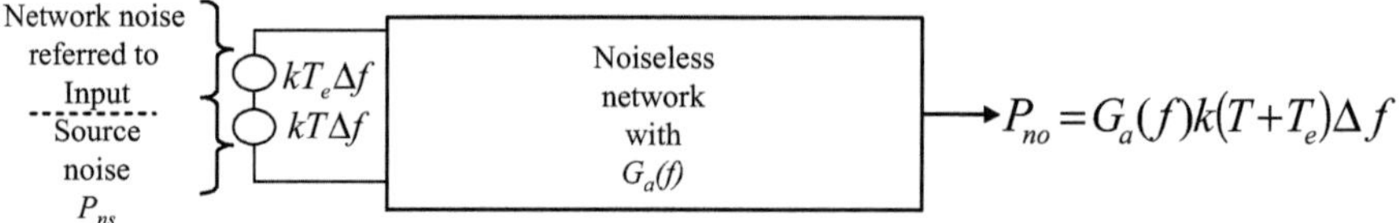

Fig. 5.7 Noisy network represented by its noise referred to the input plus noiseless network

If the network is noisy, it will produce additional noise power, P_{ne}, at the output, Fig. 5.6.

With the same input as before, the output noise power will be:

$$P_{no} = G_a(f)kT\Delta f + P_{ne} \tag{5.17}$$

Now, replacing the noisy network with a noiseless network having the same available power gain $G_a(f)$, and accounting for the output noise P_{ne} by means of an extra noise source on the input side, Fig. 5.7, the temperature T_e of this extra source is adjusted to produce P_{ne} at the output,

$$T_e = \frac{P_{ne}}{G_a(f)k\Delta f} \tag{5.18}$$

and

$$P_{no} = G_a(f)k(T + T_e)\Delta f \tag{5.19}$$

The value T_e is called the *effective input noise temperature* of the network [1, 2]. The concept of effective temperature becomes useful when analyzing a cascade of networks, each one of them characterized by its own T_e. This is discussed next.

5.5.4   Overall Noise Figure of Cascaded Networks

In general, the overall noise figure is evaluated for a bandwidth B that is the BW of the overall system, rather than that of individual stages, Fig. 5.8. If the source temperature is assumed to be T_0, which is required for the standard definition of *NF*, the output noise power of a single stage in a small frequency band is [1, 2]:

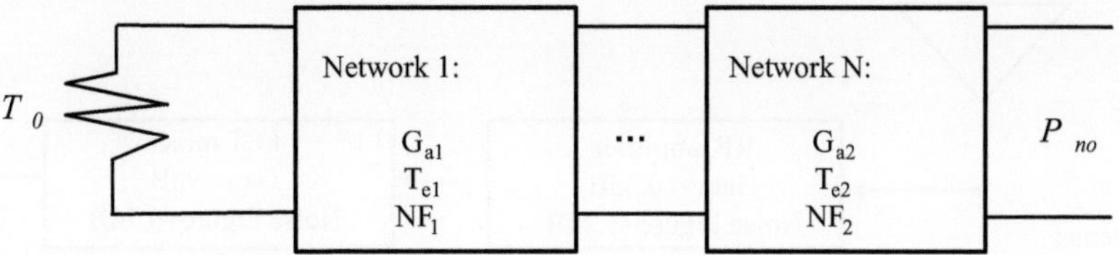

Fig. 5.8 Cascaded networks

$$P_{no} = G_a(f)k(T + T_e)\Delta f \tag{5.20}$$

In terms of the noise figure, we have,

$$P_{no} = NF \cdot G_a(f)kT_0\Delta f \tag{5.21}$$

Solving these two equations for NF gives,

$$NF = \frac{T_0 + T_e}{T_0} \tag{5.22}$$

or

$$T_e = T_0(NF - 1) \tag{5.23}$$

Thus, the noise figure, referred to a standard-temperature, may be expressed in terms of the *equivalent noise temperature* of the network, and vice versa. Therefore, the available output noise power of two cascaded networks, as shown in Fig. 5.8, is found by the use of:

$$P_{no} = G_a(f)k(T + T_e)\Delta f \tag{5.24}$$

as follows,

$$P_{no} = G_{a1}G_{a2}kT_0\Delta f + G_{a1}G_{a2}kT_{e1}\Delta f + G_{a2}kT_{e2}\Delta f \tag{5.25}$$

where the first term, to the right of the equal sign, is due to the source T_0, the second term is due to noise in the first network, and the third term is due to noise in the second network. A comparison of this relationship (5.25) with (5.18) and (5.19), repeated here for convenience,

$$T_e = \frac{P_{ne}}{G_a(f)k\Delta f} \tag{5.18}$$

and

$$P_{no} = G_a(f)k(T + T_e)\Delta f \tag{5.19}$$

leads to an expression for the effective input temperature of the two networks in cascade,

$$T_{e1,2} = \frac{kG_{a2}(G_{a1}T_{e1} + T_{e2})\Delta f}{G_{a1}G_{a2}k\Delta f} = T_{e1} + \frac{T_{e2}}{G_{a1}} \tag{5.26}$$

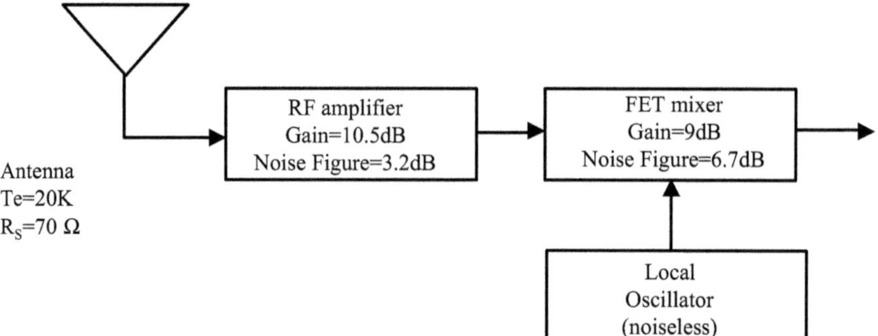

Fig. 5.9 Simplified receiver front-end

Here $T_{e1,2}$ is the effective input temperature that accounts for all of the output noise introduced by the two noisy networks in cascade. For N noisy networks in cascade, the effective temperature is [1, 2],

$$T_{e1,N} = T_{e1} + \frac{T_{e2}}{G_{a1}} + ... + \frac{T_{eN}}{G_{a1}G_{a2}...G_{a(N-1)}}$$

(5.27)

Introduction of the network noise figures by use of $T_e = T_0(NF - 1)$, leads to the following expression for the overall noise figure [2]:

$$NF_{1,N} = NF_1 + \frac{NF_2 - 1}{G_{a1}} + ... + \frac{NF_N - 1}{G_{a1}G_{a2}...G_{a(N-1)}}$$

(5.28)

Clearly, the first stage in the system has the predominant effect on the overall NF, unless G_{a1} is small or NF_2 is large.

When the input noise temperature is *not* 290 K, a true measure of the SNR degradation is the actual noise figure:

$$NF_{act} = \frac{T_S + T_e}{T_S}$$

(5.29)

in which $T_S \neq T_0$. This value is related to the "standard" noise figure evaluated for $T_S = T_0$ by the relation,

$$NF_{act} = 1 + (NF - 1)\left(\frac{T_0}{T_S}\right)$$

(5.30)

Example 1 Calculate the effective temperatures and Noise Figures for a Receiver Front End (Fig. 5.9)

In this example the following is assumed: (1) The antenna has a radiation resistance R_S and an effective temperature of 20 K due to external radiation; (2) The noise contributed by the LO is assumed to be negligible. Then, from,

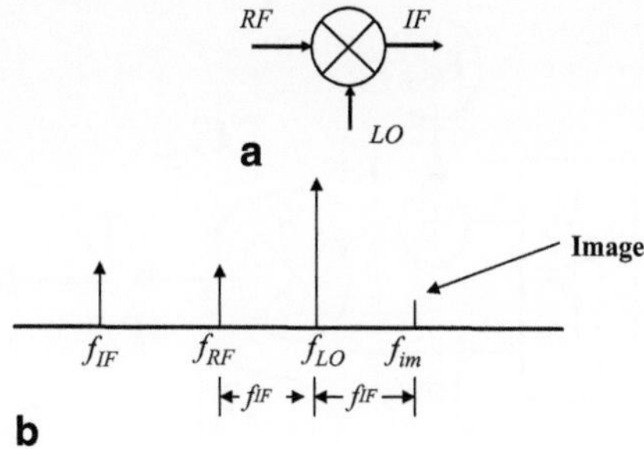

Fig. 5.10 **a** Mixer symbol.
b Realistic mixer spectrum

$$T_e = T_0(NF-1) \rightarrow T_{e1} = 290(2.04-1) = 302K \text{ and } T_{e2} = 290(4.47-1) = 1005.4K.$$

Also, from,

$$T_{e1,2} = T_{e1} + \frac{T_{e2}}{G_{a1}} \rightarrow T_r = T_{e1} + \frac{T_{e2}}{G_{a1}} = 302 + \frac{1005.4}{10} = 402.54K$$

The overall noise figure is:

$$NF = T_r / T_0 + 1 = 402.54 / 290 + 1 = 2.39$$

or

$$NF = NF_1 + (NF_2 - 1)/G_{a1} = 2.04 + 3.47/10 = 2.38 \text{ or } 3.78dB.$$ Since the antenna
temperature was 20 K (not 290 K) the actual noise figure is:

$$NF_{act} = 1 + (NF-1)\left(\frac{T_0}{T_S}\right) \rightarrow NF_{act} = 1 + (2.38-1)\left(\frac{290}{20}\right) = 21.01 \quad \text{or} \quad 13.22 \, dB$$

5.5.5 *Mixer Noise Figure*

The noise figure of a mixer is defined as the SNR at the input (RF) port divided
by the SNR at the output (IF) port [4]. However, while, ideally, a mixer would
translate only the RF frequency into the IF, in reality, the signal at the image frequency, Fig. 5.10, is also translated into the IF, thus we have the scenarios depicted
in Fig. 5.11.

Therefore, two definitions of mixer noise temperature exist [4]. Once is the Single-
Sideband (SSB) Temperature which, from an inspection of Fig. 5.11a, gives the
output noise temperature T_L as [4]:

$$T_L = (T_S + T_{SSB})G_{c\_RF} + T_S G_{c\_im} \tag{5.31}$$

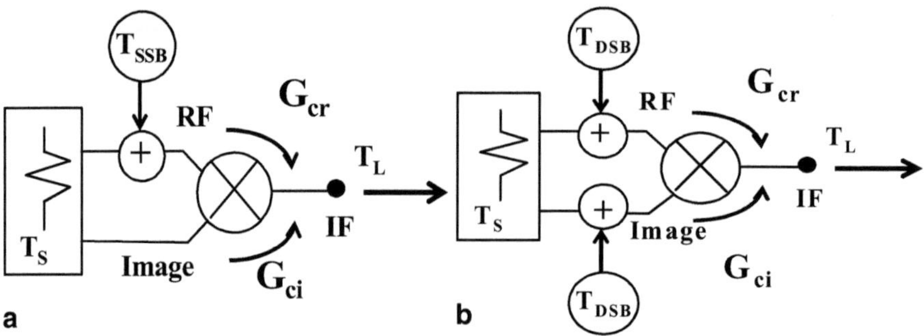

Fig. 5.11 Concept of **a** Single-Sideband and **b** Double-Sideband Noise Scenarios insert "*After* [4]"

where G_{c_RF} and G_{c_im} are the conversion gains from RF and image ports, respectively, to the IF, and T_{SSB} is the SSB noise temperature, and T_S is the input source temperature,

$$T_{SSB} = \frac{T_L - T_S(G_{c\_RF} + G_{c\_im})}{G_{c\_RF}} \qquad (5.32)$$

The other is the Double-Sideband (DSB) Noise Temperature which, similarly, from an inspection of Fig. 5.11b, gives the output noise temperature T_L as [4]:

$$T_L = (T_S + T_{DSB})G_{c\_RF} + (T_S + T_{DSB})G_{c\_im} \qquad (5.33)$$

and we obtain:

$$T_{DSB} = \frac{T_L - T_S(G_{c\_RF} + G_{c\_im})}{G_{c\_RF} + G_{c\_im}} \qquad (5.34)$$

When $G_{c\_RF} = G_{c\_im}$, a situation found often, $T_{SSB} = 2T_{DSB}$ exactly. Since noise figure is fundamentally a quantity that characterizes two-ports, applying it to a three-port component, like a mixer, creates an issue due to the ambiguity caused by having to deal with the image-frequency termination [4]. To clarify this issue, the IEEE has advanced the following definition: "For heterodyne systems, [the output noise engendered by the input termination] includes only that noise from the input termination which appears in the output via the principal-frequency transformation of the system, and does not include spurious contributions such as those from an image frequency transformation" [4].

Two interpretations have been given to the above statement:

1. The source-termination noise at frequencies other than RF should not be included when calculating the noise at the output caused by the source. This ignores the termination noise at the image, and yields the output noise temperature as $G_{c\_}$

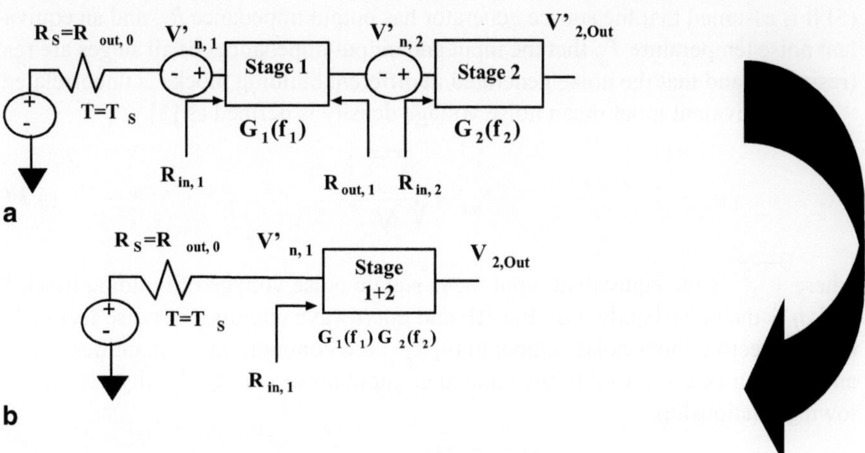

Fig. 5.12 Models for noise calculation

$_{RF}(T_{SSB} + T_0)$ and the noise from the termination alone as $G_{c\_RF}\, T_0$. The noise figure is then the ratio of these quantities,

$$NF_{SSB1} = 1 + \frac{T_{SSB}}{T_0} \tag{5.35}$$

2. A second option is to include the image noise. In this case, the image noise is treated as a noise source within the mixer, and the output temperature becomes $G_{c\_RF}(T_{SSB} + T_0) + G_{c\_im}T_0$. Assuming the conversion gains to be equal, one obtains:

$$NF_{SSB2} = 2 + \frac{T_{SSB}}{T_0} \tag{5.36}$$

This expression implies that a noiseless mixer ($T_{SSB} = 0$) will have a noise figure of 3 dB. The definition NF_{DSB} is useful in direct-conversion receivers where the image is the signal itself.

5.5.6   *Noise Calculation in Baseband*

In this sub-section, the noise scenario illustrated below is considered, following [5] closely Fig 5.12.

In modeling the cascaded system, each building block is characterized by [5]: (1) $V_{n,k}$, an equivalent input noise voltage density in units of V/Hz$^{1/2}$; (2) $R_{in, k}$, representing a noiseless input resistance; (3) $R_{out, k}$, representing a noiseless output resistance; (4) The gain of building block k, assumed to be loaded, is represented by a real voltage gain $G_k(f_k)$, where f_k denotes its input-referred pass-band frequency.

(5) It is assumed that the source generator has output impedance R_S, and an equivalent noise temperature T_S, that the input and output impedances of all stages are real (resistive), and that the noise generated in different building blocks is uncorrelated.

The equivalent input mean noise voltage density is defined as [5]:

$$V_{n,k} = \sqrt{\frac{\overline{v_{n,k}^2}}{\Delta f}} \tag{5.37}$$

where $\overline{v_{n,k}^2}$ is the equivalent input mean square noise voltage of building block k, and Δf is the noise bandwidth. For RF and microwave circuits, the noise factor, F_k, or the effective input noise temperature, $T_{e,k}$, are common noise quantities [5]. F_k and $T_{e,k}$ can be converted to the equivalent input noise voltage density via the following relationship,

$$V_{n,k} = \sqrt{4k_B T_{e,k} R_{out,k-1}} \longleftrightarrow T_{e,k} = T_0(F_k - 1) \tag{5.38}$$

where $R_{out,k-1}$ is the output resistance of the preceding stage, k_B is Boltzmann's constant, and T_0 is the standard reference temperature (290 K). It should be stressed that for mixers, the double side-band (DSB) noise factor must be used. The noise from the source, e.g., the antenna, is included in the first noise generator, $V_{n,1}$, by the following relation:

$$V_{n,1} = \sqrt{V_{n,1}'^2 + 4k_B T_S R_S} \tag{5.39}$$

where $V_{n,1}'$ represents the noise from the first stage, R_S is the source impedance, and T_S is the equivalent noise temperature of the source [5].

To work with an arbitrary impedance matching between building blocks, the parameter κ_k is defined as:

$$\kappa_k = \left(\frac{R_{in,k}}{R_{in,k} + R_{out,k-1}}\right)^2 \tag{5.40}$$

where $R_{in,k}$ is the input resistance of building block k. In the case in which power-matched stages are being dealt with, κ_k equals ¼; for fully voltage matched circuits, $\kappa_k = 1$.

In the case of frequency-translating stages, e.g. mixers, determining the output spectral noise density requires additional considerations. This is rooted in the fact that, as previously mentioned, noise from the image band of the mixer adds to the output signal band, thereby increasing the noise spectral density at the output. In practical receivers, one of two cases can be assumed: (1) The image noise is sufficiently suppressed below the noise in the wanted band and therefore image noise does not contribute to the output noise; or (2) The image noise has the same level as the noise in the wanted band and they both contribute equally to the output noise [5].

To consider the possible image noise, a factor α_k is defined for each stage as:

$$\alpha_k = \frac{\text{output noise power from image band and passband}}{\text{output noise power from passband}} \tag{5.41}$$

For building blocks with no frequency translation, $\alpha_k = 1$, whereas, for homodyne receiver mixers, equal noise from both passband and 'image band' (the lower signal side-band) translates to baseband and, therefore, $\alpha_k = 2$ [5].

In super-heterodyne receivers, the common practice includes the use of an image suppression filter in front of the mixer, which effectively suppresses the image noise and, therefore, $\alpha_k = 1$. With the above definitions, the actual noise voltage density at the output of stage k across the load terminals, $V_{k,\,out}$, is determined as:

$$V_{k,out} = \sqrt{\sum_{r=1}^{k} V_{n,r}^2 \cdot \kappa_r \cdot \prod_{i=r}^{k} \alpha_i \cdot G_i^2(f_i)} \tag{5.42}$$

Notice that, when studying noise levels throughout the receiver chain, this equation does not include noise from stages $k > K$.

Including all N cascaded building blocks, the total effective noise temperature of the receiver, referred to the input, T_e, is calculated as [5],

$$T_e = \frac{1}{4k_B R_S \kappa_1} \left(\sum_{r=1}^{N} V_{n,r}^2 \cdot \kappa_r \cdot \prod_{i=1}^{r-1} \frac{1}{\alpha_i G_i^2(f_i)} \right) - T_S \tag{5.43}$$

Since each of the terms in the sum of T_e represents the effective noise temperature of each stage, we can inspect them to assess which are the main noise contributors of the receiver. Upon calculating the total output noise voltage density, $V_{N,\,out}$ by (5.42), T_e may also be determined from:

$$T_e = \frac{V_{N,out}^2}{4k_B R_S \kappa_1} \cdot \prod_{i=1}^{N} \frac{1}{\alpha_i G_i^2(f_i)} - T_S \tag{5.44}$$

Using T_e, the noise factor F of the overall cascade is calculated by the well-known relation [4]:

$$NF = 1 + \frac{T_e}{T_0} \tag{5.45}$$

5.6   Nonlinearity and Time Variance

We begin this sub-section by reviewing the topic of nonlinearity, which was first introduced in Chap. 3 in the context of characterizing amplifiers. The goal here is to see how the nonlinearity of building blocks in a cascade making up a system impacts the overall system.

The input-output transfer curve of a building block, e.g., an amplifier, may be represented as,

$$y(t) = F(x(t)) = \sum_{n=1}^{\infty} a_n x^n(t) \cong \sum_{n=1}^{N} a_n x^n(t) \tag{5.46}$$

which, retaining the first four terms, may be written as,

$$y(t) = a_0 + a_1 x(t) + a_2 x^2(t) + a_3 x^3(t) \tag{5.47}$$

In this equation, the first two terms capture the linear part, and the third and fourth terms capture the nonlinear part; this nonlinearity gives rise to the problems of gain compression and desensitization.

5.6.1 Gain Compression and Desensitization

Consider the presence of the following signals at the input of an amplifier,

$$S_d(t) = A_d \cos 2\pi f_0 t \tag{5.48}$$

$$S_{I1}(t) = A_{I1} \cos 2\pi f_1 t \tag{5.49}$$

$$S_{I2}(t) = 0 \tag{5.50}$$

where S_d, S_{I1} and S_{I2} represent the desired, and two possible interfering signals, the latter with zero amplitude. Then, substituting these into (5.47) one gets,

$$y(t) = a_1 \left[S_d(t) + S_{I1}(t) + S_{I2} \right] + a_2 \left[S_d(t) + S_{I1}(t) + S_{I2}(t) \right]^2$$
$$+ a_3 \left[S_d(t) + S_{I1}(t) + S_{I2}(t) \right]^3 + \dots \tag{5.51}$$

This, upon insertion of (5.47)–(5.50), yields,

$$y(t) = a_1 \left[A_d \cos 2\pi f_0 t + A_{I1} \cos 2\pi f_1 t \right]$$
$$+ a_2 \left[A_d^2 \cos^2 2\pi f_0 t + A_{I1}^2 \cos^2 2\pi f_1 t + 2 A_d A_{I1} \cos 2\pi f_0 t \cos 2\pi f_1 t \right]$$
$$+ a_3 \left[A_d^3 \cos^3 2\pi f_0 t + A_{I1}^3 \cos^3 2\pi f_1 t + 3 A_d^2 A_{I1} \cos^2 2\pi f_0 t \cos 2\pi f_1 t \right.$$
$$\left. + 3 A_d A_{I1}^2 \cos 2\pi f_0 t \cos^2 2\pi f_1 t \right] + \dots \tag{5.52}$$

which may be simplified as,

$$y(t) = a_1 A_d \left[1 + \frac{3a_3}{4a_1} A_d^2 + \frac{3a_3}{2a_1} A_{I1}^2 \right] \cos 2\pi f_0 t + \dots \tag{5.53}$$

Fig. 5.13 Description of 1 dB compression point

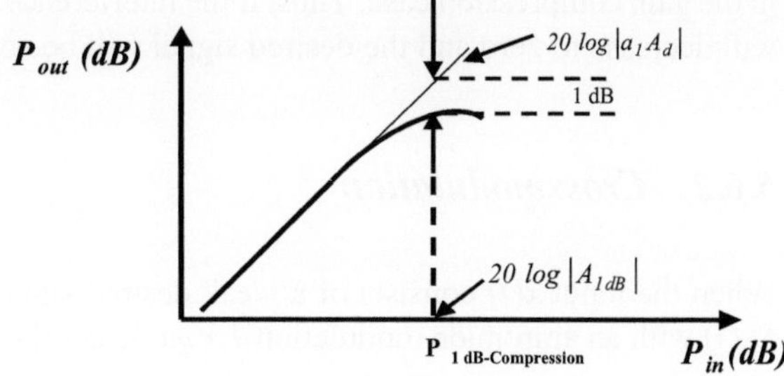

Now, when the *interference is absent*, and the desired signal amplitude, $A_d \ll 1$, the *small signal gain* is equal to a_1, since all the other terms above will be negligibly small for a mildly nonlinear system, when compared to the first term $a_1 A_d \cos 2\pi f_0 t$. However, as the amplitude of the desired signal increases, the gains, as determined by the term within brackets, will vary with the input signal level. In particular, the second term in the brackets, $3a_3 A_d^2 / 4a_1$, gradually becomes significant. Then, *if the sign of a_3 is opposite that of a_1*, the output, $y(t)$, will be smaller than that predicted by linear theory giving rise to the phenomenon called *gain compression*. The compressive gain G_c in decibels (dB) is given by:

$$G_c = 20 \log \left| a_1 \left(1 + \frac{3a_3}{4a_1} A_d^3 \right) \right| \tag{5.54}$$

Clearly, given opposite signs in a_3 *and* a_1 the compressive gain decreases with increasing A_d. This gain compression is characterized by the signal level at which $A_d = A_{-1}$, i.e., at which the gain is 1 dB lower than the small signal gain, which is referred to as the *1 dB compression point* of the amplifier. The value of A_{-1} is found by setting,

$$G_c = 20 \log \left| a_1 \left(1 + \frac{3a_3}{4a_1} A_{-1}^3 \right) \right| = -1 \tag{5.55}$$

and solving for A_{-1}, this gives,

$$A_{-1} = \sqrt{\left(1 - 10^{-\frac{1}{20}} \right) \frac{4}{3} \left| \frac{a_1}{a_3} \right|} = \sqrt{0.145 \left| \frac{a_1}{a_3} \right|} \tag{5.56}$$

The concept of 1 dB compression point is illustrated in Fig. 5.13.

In the presence of interference, the desired output signal will also decrease when the amplitude A_{I1} of the interferer increases, if $a_3 < 0$, see (5.53). In particular, the interference is responsible for a decrease in gain by the amount $3a_3 A_{I1}^2 / 2a_1$. This nonlinear effect is called *desensitization*, and manifests itself in that, when it occurs, the gain experienced by weak desired signal will be very small. In fact, in the presence of a strong interference, the gain decreases at *twice the rate* as compared to that in the gain compression case. Thus, if the interference is sufficiently strong, the gain will decrease to zero and the desired signal will be completely *blocked.*

5.6.2   Crossmodulation

When the input $x(t)$ consists of a weak desired signal $S_d(t)$ and a strong interferer $S_{I1}(t)$ with an amplitude modulation $1 + m(t)$, as follows:

$$\left. \begin{array}{l} S_d(t) = A_d \cos 2\pi\, f_0 t \\ S_{I1}(t) = A_{I1}[1 + m(t)] \cos 2\pi\, f_1 t \\ S_{I2}(t) = 0 \end{array} \right\} \tag{5.57}$$

the output, $y(t)$, of a nonlinear system is then:

$$y(t) = a_1 A_d \left[1 + \frac{3a_3}{4a_1} A_d^2 + \frac{3a_3}{2a_1} A_{I1}^2 \left[1 + m^2(t) + 2m(t) \right] \right] \cos 2\pi f_0 t + \dots \tag{5.58}$$

This expression contains desensitization and compression terms and two new terms, namely, $(3a_3/2a_1)A_{I1}^2 m^2(t)$, and $(3a_3/a_1)A_{I1}^2 m(t)$. These new terms mean that the amplitude modulation on the strong interferer is now transferred to the desired signal through the interaction with the nonlinearity. This nonlinear phenomenon is referred to as *crossmodulation.*

5.6.3   Intermodulation

Consider the case when there is more than one interferer, $S_{I1}(t)$ and $S_{I2}(t)$ (5.59) accompanying the desired signal $S_d(t)$ in the input $x(t)$.

$$\left. \begin{array}{l} S_d(t) = A_d \cos 2\pi\, f_0 t \\ S_{I1}(t) = A_{I1} \cos 2\pi\, f_1 t \\ S_{I2}(t) = A_{I2} \cos 2\pi\, f_2 t \end{array} \right\} \tag{5.59}$$

Then, the output becomes,

$$\begin{aligned} y(t) = {}& a_1 A_d \left[1 + \frac{3a_3}{4a_1} A_d^2 + \frac{3a_3}{2a_2} \left(A_{I1}^2 + A_{I2}^2 \right) \right] \cos 2\pi f_0 t \\ & + a_2 A_{I1} A_{I2} \left[\cos 2\pi (f_1 + f_2)t + \cos 2\pi (f_1 - f_2)t \right] \\ & + \frac{3}{4} a_3 \left[A_{I1}^2 A_{I2} \cos 2\pi (2f_1 \pm f_2)t + A_{I1} A_{I2}^2 \cos 2\pi (2f_2 \pm f_1)t \right] + \dots \end{aligned} \tag{5.60}$$

In (5.60), the terms with frequencies $(f_1 + f_2)$, are called *second-order intermodulation products,* whereas the terms with frequencies $(2f_1 \pm f_2)$ and $(2f_2 \pm f_1)$, are

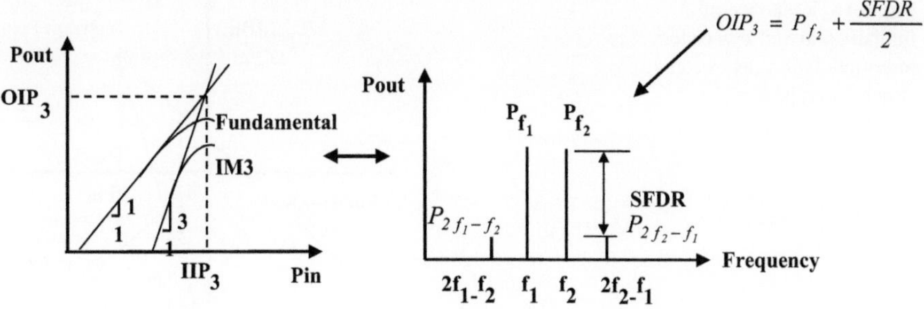

Fig. 5.14 Power transfer and power spectral relationships illustrating SFDR concept

Fig. 5.15 Cascaded building blocks/stages making up a system

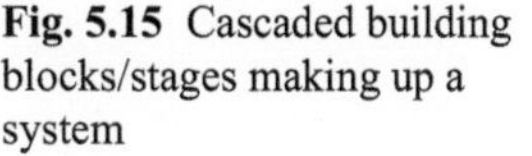
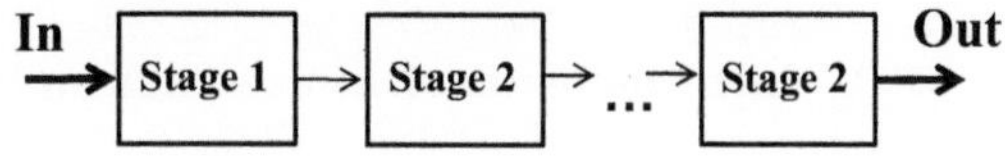

called *third-order intermodulation products*. The latter frequencies are easily seen in Fig. 5.14.

5.6.4 Cascaded Nonlinear Stages

The issue addressed in this sub-section is this: How do we characterize the effects of nonlinearity in cascaded nonlinear stages, Fig. 5.15?

5.6.4.1 Second-Order Input Intercept Point

The analysis of intermodulation (IM) distortion is facilitated by the concept of *intercept point*, which provides a convenient and measurable quantity. In the treatment of IM distortion, three general issues must be defined, namely:

1. The type and frequency of the interfering signals that cause the IM distortion;
2. The frequency location of the IM distortion: For *odd-order* distortion, the products of concern most often fall on/close to the fundamental frequency, whereas, *for even-order* IM distortion, the low-frequency products are often considered, in particular, for low-IF and homodyne receivers where most signal processing is conducted at baseband;
3. The intercept point (IP) is related to the above scenario: The IP must be measured with correct frequency and type of interferer, as well as the correct source and load impedances.

Notice that intercept points are small-signal nonlinearity measures, which cannot be applied under large-signal conditions, i.e. near compression. Other key stage parameters are the pass-band gain, the stage selectivity at the interfering frequency, and the stage selectivity at the distortion frequency [5, 6].

Fig. 5.16 Response of fundamental and *mth* order intermodulation distortion versus input power

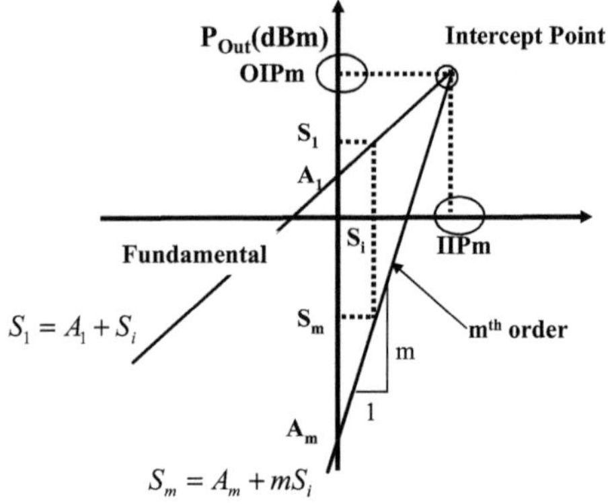

5.6.4.2   Intermodulation Distortion Formulas

As shown in Chap. 3, in the weak and memoryless nonlinear case, a power series may be used to model the nonlinearity of a device such as an amplifier or a mixer. If the input signal and output signal power of the device is P_i and P_o, respectively, then the output of the nonlinear device can be represented by its input as follows [6]:

$$P_o = \sum_{m=1}^{n} P_m = \sum_{m=1}^{n} \alpha_m P_i^m \tag{5.61}$$

where α_m ($m = 1, 2, ..., n$) is the power gain when $m = 1$, and the *nonlinear gain coefficients,* when $m \neq 1$, and P_m is the *mth* order distortion power of the total output power P_o, and it is related to the input signal power P_i as [6]:

$$P_m = \alpha_m P_i^m \quad (m = 1 \cdots n) \tag{5.62}$$

It can be seen that the fundamental and *mth* order terms, namely,

$$P_1 = \alpha_1 P_i \tag{5.63}$$

and

$$P_m = \alpha_m P_i^m \tag{5.64}$$

are special in that (5.63) describes a linear relationship between the output and input, and (5.64) represents a generic high-order distortion. Using *decibels* (dB), (5.63) and (5.64) may be expressed as follows [6]:

$$S_1 = A_1 + S_i \tag{5.65}$$

$$S_m = A_m + mS_i \tag{5.66}$$

where $S_1 = 10\log P_1$, $A_1 = 10\log \alpha_1$, $S_i = 10\log P_i$, $S_m = 10\log P_m$, $A_m = 10\log \alpha_m$, and $mS_i = 10\log P_i^m$. The expressions (5.65) and (5.66) are indicated in Fig. 5.16.

From Fig. 5.16 and the above equations one is able to obtain two expressions of the output power at the intercept point, *OIPm*, when the input power equals *IIPm*, namely, [6],

$$OIP_m = A_1 + IIP_m \tag{5.67}$$

$$OIP_m = A_m + m \cdot IIP_m \tag{5.68}$$

From (5.65) through (5.68), we get,

$$(m-1)(IIP_m - S_i) = S_1 - S_m \tag{5.69}$$

Now, if some notations that are more familiar in RF system design are employed, namely, $S_i = I$, which assumes that the input is an interference, and S_m, the *mth*-order distortion, is represented by the sum of the equivalent level at the device input, IMD_m, and the linear gain of the device A_1, or,

$$S_m = IMD_m + A_1 \tag{5.70}$$

Then, after considering $S_1 = I + A_1$, one finally obtains the general *IMD* formula,

$$(m-1)(IIP_m - S_i) = S_1 - S_m \tag{5.71}$$

Taking,

$$(m-1)(IIP_m - I) = I - IMD_m \tag{5.72}$$

we get,

$$IMD_m = mI - (m-1)IIP_m \tag{5.73}$$

from which,

$$IIP_m = \frac{mI - IMD_m}{m-1} \tag{5.74}$$

For $m = 2$ and *3*, one has, respectively,

$$IIP_2 = 2I - IMD_2 \tag{5.75}$$

for the *second-order intercept point*, and

$$IIP_3 = \frac{1}{2}(3I - IMD_3) \tag{5.76}$$

for the *third-order intercept point*. Figure 5.17 is an example of data from measured amplifier.

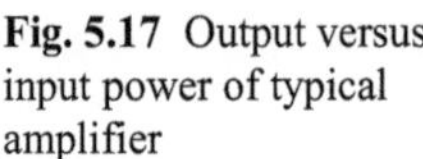

Fig. 5.17 Output versus input power of typical amplifier

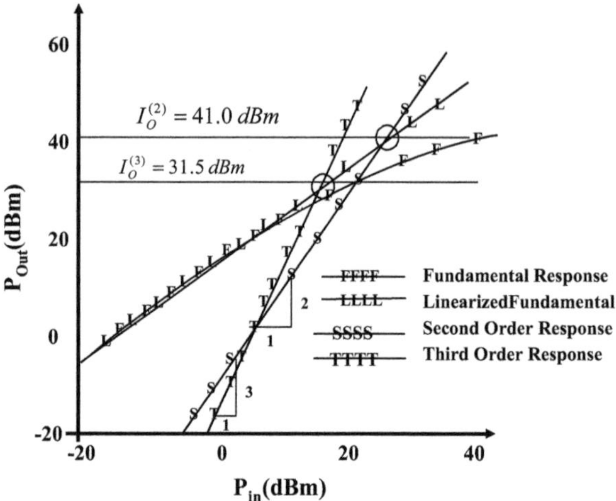

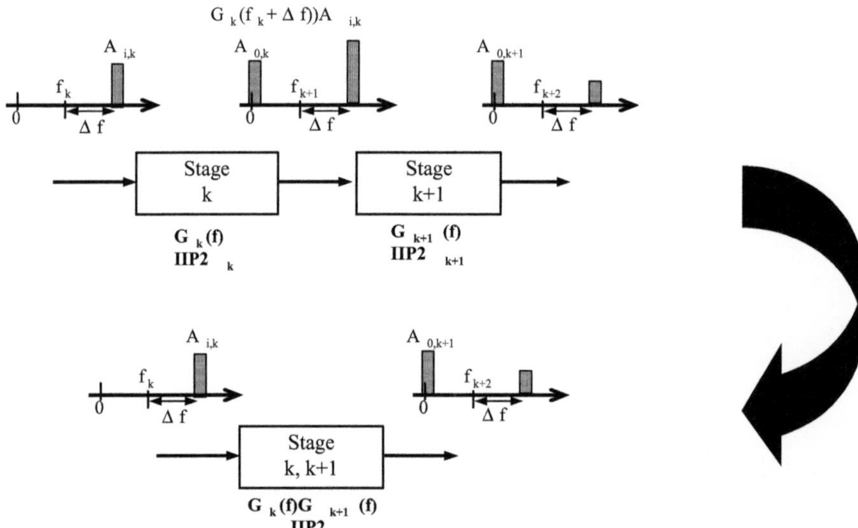

Fig. 5.18 *IIP2* of individual cascade stages is combined into that of a single stage

5.6.4.3   Cascade Second-order Input Intercept Point

The goal of this sub-section is to describe how to calculate the second-order input intercept point, *IIP2* for a cascade of receiver building blocks. In doing so it is assumed that the *IIP2*, gain, and selectivity for the individual stages is known, and that, as shown previously, the *IIP2*, represents a fictitious input level at which the extrapolated desired response intersects with the extrapolated second-order *IM* product of interest [5]. The scenario in question is depicted in Fig. 5.18.

Utilizing the notation in Fig. 5.18, it can be shown that the rms amplitude $A_{0,k}$, of the second-order *IM* product at the output of stage k, is given by [5]:

$$A_{o,k} = A_{i,k}^2 \cdot \frac{G_k}{IIP2_k} \tag{5.77}$$

where $A_{i,k}$ is the interferer input rms amplitude, $G_k(f_k)$ is the desired signal gain, and $IIP2_k$ is the second-order input interception point. At the output of two cascaded stages k and $k+1$, there is a total second-order *IM* combination given by [5]:

$$A_{o,k+1} = G_{k+1}(f_{im}) \cdot A_{i,k}^2 \frac{G_k(f_k)}{IIP2_k}$$
$$+ A_{i,k}^2 G_k^2(f_k + \Delta f) \cdot \frac{G_{k+1}(f_{k+1})}{IIP2_{k+1}} \tag{5.78}$$

In (5.78) the first term represents the *IM* product generated in building block k and passing to towards the output through building block $k+1$. The second term is the *IM* product generated in building block $k+1$ itself. (5.78) assumes the worst-case scenario where the *IM* products generated in blocks k and $k+1$ add in phase at the output of block $k+1$, i.e., resulting in a maximum. As shown in the figure, $G_{k+1}(f_{im})$ is the gain of block $k+1$ with relative to the *IM* product [5].

In (5.78), it is also assumed that the gain in block k, pertaining to two possible interfering signals, can be represented by a single quantity $G_k(f_k + \Delta f)$. This assumption is appropriate in most practical cases in which the two interfering signals are close in frequency, i.e., virtually no frequency dependence is manifested. One can then compute the total second-order IM combination at the output of two cascaded blocks k and $k+1$ by [5]:

$$A_{o,k+1} = G_{k+1}(f_{im}) \cdot A_{i,k}^2 \frac{G_k(f_k)}{IIP2_k}$$
$$+ A_{i,k}^2 G_k^2(f_k + \Delta f) \cdot \frac{G_{k+1}(f_{k+1})}{IIP2_{k+1}} \tag{5.79}$$

The IIP2 of the cascaded stages k and $k+1$ can then be shown to be given by:

$$\frac{1}{IIP2} = \frac{G_{k+1}(f_{im})}{G_{k+1}(f_{k+1}) \cdot IIP2_k} + \frac{G_k^2(f_k + \Delta f)}{G_k(f_k) \cdot IIP2_{k+1}} \tag{5.80}$$

In general, the cascade of N *stages* gives a total second order input intercept point of [5],

$$\frac{1}{IIP2} = \sum_{k=1}^{N} \frac{I_k \cdot J_k}{IIP2_k} \tag{5.81}$$

where

$$I_k = \prod_{r=1}^{k-1} \frac{G_r^2(f_r + \Delta f)}{G_r(f_r)} \tag{5.82}$$

and

$$J_k = \prod_{r=k+1}^{N} \frac{G_r(f_{im})}{G_r(f_r)} \tag{5.83}$$

5.6.4.4 Two-tone IIP2

An *IIP2* measure that is commonly employed is the *two-tone IIP2*. This is obtained, both in measurement and simulation, by applying at the input two closely-spaced sinusoidal tones with equal amplitudes. The second-order *IM* product at the output is a low frequency signal with a frequency equal to the spacing between the two sinusoidal tones f_1 and f_2. Let's assume a memoryless nonlinear system described by the following transfer function:

$$y(t) = a_0 \cdot x^0(t) + a_1 \cdot x^1(t) + a_2 \cdot x^2(t) + a_3 \cdot x^3(t) + \ldots \tag{5.84}$$

where $x(t)$ is the input voltage and $y(t)$ is the output voltage, and the factor a_1 represents the voltage gain. Then, the two-tone input signal, $x_{tt}(t)$, where both tones have rms amplitude A_{tt}, is given by:

$$x_{tt}(t) = \sqrt{2} \cdot A_{tt} \cdot (\cos 2\pi f_1 t + \cos 2\pi f_2 t) \tag{5.85}$$

Inserting (5.85) into (5.84), the output signal is a sum of the fundamental tones and *IM* products, with the low-frequency second-order *IM* product at the output being:

$$y_{im,tt}(t) = 2 \cdot a_2 \cdot A_{tt}^2 \cdot \cos(2\pi (f_1 - f_2) t) \tag{5.86}$$

The two-tone second-order input intercept point $IIP2_{tt}$ is then defined as the input rms amplitude that results in equal output amplitudes of a desired signal and the IM product $y_{im,\,tt}$. Applying a desired input signal with an rms amplitude of $IIP2_{tt}$, gives rise to an output rms amplitude of $|a_1| \cdot IIP2_{tt}$. Therefore [5],

$$|a_1| \cdot IIP2_{tt} = \frac{2 \cdot |a_2| \cdot IIP2_{tt}^2}{\sqrt{2}} \tag{5.87}$$

from where,

$$IIP2_{tt} = \frac{|a_1|}{\sqrt{2} \cdot |a_2|} \tag{5.88}$$

5.6.4.5 Single-Tone IIP2

The IIP2 may also be derived from a single-tone [5]. In this case, a single sinusoid, or potentially phase-modulated signal, is applied at the input and the resulting second-order output *IM* product is located at DC. Assuming the transfer function given previously, namely,

$$y(t) = a_0 \cdot x^0(t) + a_1 \cdot x^1(t) + a_2 \cdot x^2(t) + a_3 \cdot x^3(t) + \ldots \tag{5.89}$$

and an input signal of,

$$x_{st}(t) = \sqrt{2} \cdot A_{st} \cdot \cos\left(2\pi f_1 t + \phi(t)\right) \tag{5.90}$$

the resulting *IM* product at DC is given as,

$$y_{im,st}(t) = a_2 \cdot A_{st}^2 \tag{5.91}$$

Using the same approach as above, the single-tone IIP2 becomes,

$$IIP2_{st} = \frac{|a_1|}{|a_2|} \tag{5.92}$$

The relation between the single-tone and the two-tone intercept point is thus given by,

$$IIP2_{st} = \sqrt{2} \cdot IIP2_{tt} \tag{5.93}$$

This is valid for memoryless circuits or circuits with DC coupling [5].

5.6.4.6   Time-Varying-Envelope Signal IIP2

Systems requiring high spectral efficiency usually employ modulation approaches that involve a time-varying signal envelope. For such signals, it is not possible to apply directly the two-tone and single-tone *IIP2* measures to predict the resulting second-order *IM* product [5].

   Instead, a modulated signal represented in vector form is utilized. Such a signal is given by,

$$x_{IQ}(t) = \sqrt{2} \cdot A_{IQ} \cdot \left(I(t) \cdot \cos 2\pi f_1 t + Q(t) \cdot \sin 2\pi f_1 t\right) \tag{5.94}$$

which assumes that $I(t)$ and $Q(t)$ are scaled such that $\sqrt{\langle I^2(t)\rangle_t + \langle Q^2(t)\rangle_t} = 1$, and $\langle \cdot \rangle_t$ represents time averaged mean; the rms voltage of $x_{IQ}(t)$ is simply A_{IQ}. By applying the signal $x_{IQ}(t)$ to the nonlinear transfer function,

$$y(t) = a_0 \cdot x^0(t) + a_1 \cdot x^1(t) + a_2 \cdot x^2(t) + a_3 \cdot x^3(t) + \ldots \tag{5.95}$$

the low-frequency *IM* product at the output is given by,

$$y_{im,IQ}(t) = a_2 \cdot A_{IQ}^2 \, I^2(t) + Q^2(t) \tag{5.96}$$

and the rms amplitude of this *IM* product is given by,

$$A_{im.IQ} = |a_2| \cdot A_{IQ}^2 \cdot K_{IQ} \tag{5.97}$$

and,

$$K_{IQ} = \sqrt{\left\langle (I^2(t) + Q^2(t)) \right\rangle_t} \tag{5.98}$$

where K_{IQ} describes the power variation of $x_{IQ}(t)$ [5].

Using the same approach as for the two-tone $IIP2$,

$$|a_1| \cdot IIP2_{IQ} = |a_2| \cdot IIP2_{IQ}^2 K_{IQ} \tag{5.99}$$

from where,

$$IIP2_{IQ} = \frac{|a_1|}{K_{IQ} \cdot |a_2|} \tag{5.100}$$

The relation to the two-tone $IIP2$ is thus given by,

$$IIP2_{IQ} = IIP2_{tt} \cdot \frac{\sqrt{2}}{K_{IQ}} \tag{5.101}$$

The procedure for measuring $IIP2_{IQ}$ is more difficult than that for measuring $IIP2_{tt}$, since it requires a source with the correct modulation. In addition, due to the modulated input signal, simulation is also more complicated since harmonic balance simulation techniques cannot be applied; instead, time-domain or circuit-envelope techniques are needed. With proper knowledge of K_{IQ} and $IIP2_{tt}$ from measurement or simulation, (5.101) provides a straightforward conversion to $IIP2_{IQ}$.

5.6.4.7   Third-Order Input Intercept Point

The third-order input intercept point, $IIP3$, represents a fictitious input level at which the extrapolated desired response intersects with the extrapolated third-order response of interest. Due to third-order nonlinearities, two interferers at frequencies $f_k + \Delta f$ and $f_k + 2\Delta f$, result in an IM product that coincides with the desired signal frequency f_k. Therefore, this type of distortion is **not separable** from the desired signal!

In this sub-section, the description of how the $IIP3$ can be determined for a cascade of receiver building blocks/stages and how the commonly applied two-tone $IIP3$ can be determined for a memoryless nonlinearity, are addressed.

Using the notation in Fig. 5.19, the output rms amplitude of the third-order IM product generated in building block k is given by,

$$A_{o,k} = A_{1,k}^2 A_{2,k} \frac{G_k(f_k)}{IIP3_k^2} \tag{5.102}$$

where $A_{1,k}$ is the rms amplitude of the interfering signal located nearest to the passband frequency. The third-order output contribution after blocks k and $k+1$ is given by,

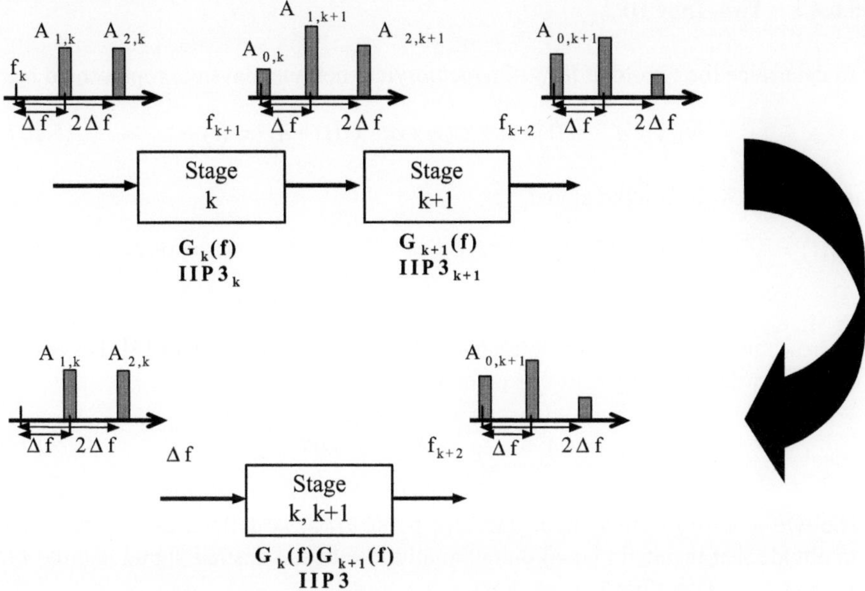

Fig. 5.19 *IIP3* of individual cascade stages is combined into that of a single stage

$$A_{o,k+1} = G_{k+1}(f_{k+1}) \cdot A_{1,k}^2 A_{2,k} \frac{G_{k+1}(f_{k+1})}{IIP3_{k+1}^2}$$

$$+ G_k^2\left(f_k + \Delta f\right) A_{1,k}^2 G_k^2\left(f_k + 2\Delta f\right) A_{2,k} \frac{G_{k+1}(f_{k+1})}{IIP3_{k+1}^2} \tag{5.103}$$

which assumes that the generated *IM* products of blocks k and $k+1$ add in phase at the output of stage $k+1$. The overall *IIP3* of the two cascaded stages k and $k+1$ can be shown to be given by,

$$\frac{1}{IIP3^2} = \frac{1}{IIP3_k^2} + \frac{G_k^2(f_k + \Delta f) \cdot G_k^2(f_k + 2\Delta f)}{G_k(f_k) \cdot IIP3_{k+1}^2} \tag{5.104}$$

In general, the total third-order intercept point for a cascade of N building blocks is given by [5],

$$\frac{1}{IIP3^2} = \sum_{k=1}^{N} \frac{H_k}{IIP3_k^2} \tag{5.105}$$

where

$$H_k = \prod_{r=1}^{k-1} \frac{G_r^2(f_r + \Delta f) \cdot G_r(f_r + 2\Delta f)}{G_r(f_r)} \tag{5.106}$$

5.6.4.8   Two-Tone IIP3

To determine the two-tone *IIP3* of a memoryless nonlinear system represented by,

$$y(t) = a_0 \cdot x^0(t) + a_1 \cdot x^1(t) + a_2 \cdot x^2(t) + a_3 \cdot x^3(t) + \dots \tag{5.107}$$

it is excited with an input signal,

$$x_{tt}(t) = \sqrt{2} \cdot A_1 \cdot \cos\left((2\pi f_0 + 2\pi \Delta f)t\right) + \sqrt{2} \cdot A_2 \cdot \cos\left((2\pi f_0 + 2 \cdot 2\pi \Delta f)t\right)$$

$$\tag{5.108}$$

where A_1 and A_2 are the rms amplitudes of the two interfering signals [5]. The resulting third-order *IM* product at f_0 is given by,

$$y_{im,tt}(t) = \frac{3 \cdot a_3}{\sqrt{2}} \cdot A_1^2 \cdot A_2 \cdot \cos 2\pi f_0 t \tag{5.109}$$

The two-tone third-order input intercept point, *IIP3*, is defined as the input rms amplitude that results in equal output amplitudes of the desired signal and the *IM* product $y_{im,\,tt}$. Therefore,

$$|a_1| \cdot IIP3 = \frac{3 \cdot |a_3| \cdot IIP3^3}{2} \tag{5.110}$$

from where,

$$IIP3_{tt} = \sqrt{\frac{2}{3} \frac{|a_1|}{|a_3|}} \tag{5.111}$$

When testing the receiver, the interferer in offset $2\Delta f$ is typically a modulated signal whereas the interferer in offset Δf is a sinusoidal. For this setup, the IIP3 given by (5.111) still applies.

5.7   Summary

In this chapter we have dealt with the topics of Noise, Nonlinearity and Time Variance. We began with a presentation of the various noise sources of importance to a receiver, in particular, Thermal Noise, and its definition, quantification and voltage and current circuit models; and Shot noise and its quantification. Then we took on the topic of Signal-to-Noise Ratio, including, available noise power, Noise Figure, Noise Temperature, and Noise Figure of Cascaded Linear Networks. This was followed by the topic of Mixer Noise Figure, in particular, Single-Sideband and Double-Sideband. We concluded this part with a discussion of the calculation of the noise at baseband for a system consisting of a cascade of building blocks. Then, we addressed the topics of Nonlinearity and Time Variance. In particular, we reviewed effects of nonlinearity, namely, Gain compression and Intermodulation, and then addressed the topics of Cascaded nonlinear stages, Second-order intercept point, Intermodulation Distortion Formulas, Cascade Second-order input intercept point (*IIP2*), Two-tone *IIP2*, One-tone *IIP2*, time-varying-envelope signal *IIP2*, third-order intercept point (*IIP3*), cascaded third-order input intercept point (*IIP3*), and two-tone *IIP3*.

References

1. Mischa Schwartz, *Information Transmission, Modulation, and Noise*, McGraw-Hill, 1970.
2. G.H. Wannier, *Statistical Physics*, Dover Publications, Inc., New York, 1987.
3. H.L. Krauss, C.W. Bostian, and F.H. Raab, *Solid State Radio Engineering*, John Wiley & Sons, Inc., 1980.
4. S.A. Maas, *Noise in Linear and Nonlinear Circuits*, Boston, Artech House (2005).
5. C.R. lversen and T.E. Kolding, "Noise and intercept point calculation for modern radio receiver planning," *IEE Pro < .-Commun., Vol. 148, No. 4, August 2001, pp. 255–259*.
6. Q. Gu, *RF System Design of Transceivers for Wireless Communications*, Springer, 2005.

Chapter 6
Sensitivity and Dynamic Range

Abstract In this chapter the topics of sensitivity and dynamic range for a receiver are addressed. We begin by introducing their definitions, the concepts of minimum detectable signal, signal-to-noise ratio, 1 dB compression point (CP 1 dB), saturation, and some relationships between input and output CP 1 dB), intermodulation distortion, thirdorder intercept point (IP3), receiver blocking, and spur-free dynamic range. The relationships between these parameters as applied to building blocks in a cascade arrangement making up a receiver, and the overall parameters for the receiver, is presented. The chapter ends with an example of how changes in the sequence of building blocks in a cascade impact the sensitivity and dynamic range of a receiver.

6.1 Introduction

The design of a communications receiver entails an iterative process which ultimately results in a compromise among conflicting performance goals [1]. This is so because, on the one hand, the smallest signals a receiver is capable of detecting and processing is limited by noise, but on the other, the largest signal the receiver can successfully process is limited by distortions arising from circuit nonlinearities. A compromise must then be reached to achieve the largest possible dynamic range, which is the difference between the largest and the smallest signals a receiver system is capable of processing.

To achieve the lowest noise performance receiver, a designer must gain, during the design process, an understanding of the fundamental causes of noise in the receiver circuits. Similarly, because large interfering signals generate distortion products that fall within the receiver passband, thus, precluding the reception of small signals and reducing message reliability, these normally result in a suboptimal large-signal performance. Further, it is found that the best large signal performance suffers from higher noise degradation, which in turn limits the weak signal reception. Thus, to get best small- and large-signal receiver performance, an optimization trade-off must be undertaken [1].

A number of fundamental receiver architectures can be chosen from, in particular, the *homodyne* (zero-IF) receiver, Fig. 6.1, and the *heterodyne* receiver, Fig. 6.2, architectures.

H. J. De Los Santos et al., *Radio Systems Engineering,*
DOI 10.1007/978-3-319-07326-2_6, © The Authors 2015

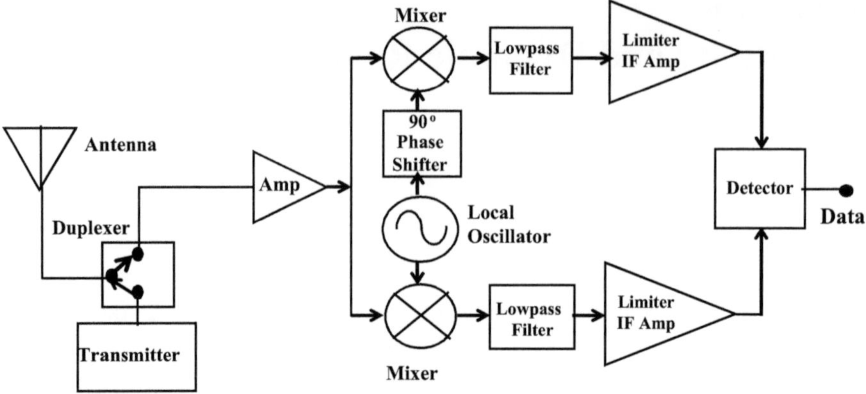

Fig. 6.1 Homodyne (zero-IF) receiver architecture

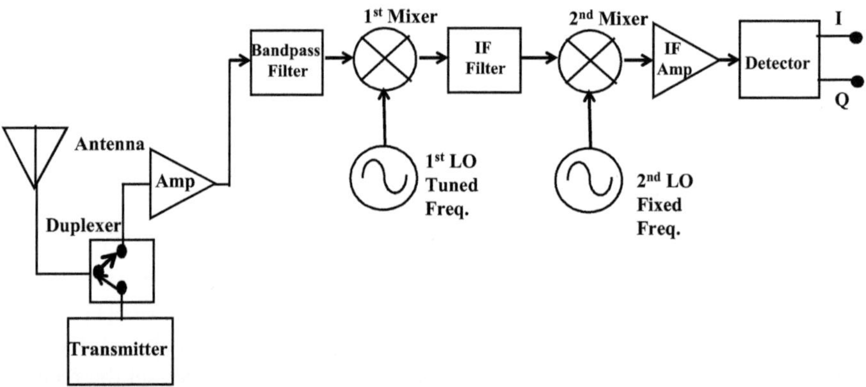

Fig. 6.2 Heterodyne receiver

In the homodyne receiver (also referred to as, direct conversion, or zero-IF, i.e., with no intermediate frequency) the desired incoming RF frequency is translated directly to baseband, where it is process and the modulation information recovered. As indicated in an earlier chapter, the baseband signal contains the range of frequencies occupied by the signal before modulation or after demodulation. These signals normally occupy frequencies well below the carrier frequencies, even approaching DC on the low end.

In the heterodyne receiver the desired incoming RF frequency is translated to one or more intermediate frequencies before demodulation. The modulation information, in turn, is recovered from the last IF frequency.

In a *dual-conversion superheterodyne* receiver, Fig. 6.2, successive mixers translate the RF signal to two IF frequencies. LO signals, tuned at a particular spacing above or below the RF signal, drive the mixer circuits. The RF and LO signals mix

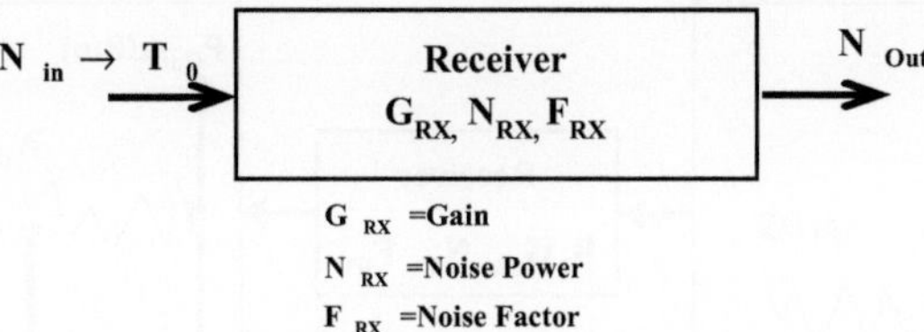

Fig. 6.3 System-level receiver parameters. F is the system's noise factor. $kT_{0\times}$(1 Hz) is the thermal noise power in a 1-Hz bandwidth at T_0, B is the noise equivalent bandwidth, G is the two-port's numeric available gain

to produce a difference frequency known as the IF frequency. The result is a dual-conversion receiver, described as such because of the two down-conversion mixers.

6.2 Sensitivity

Sensitivity is one of the most important parameters determining the overall performance of a communication system, as it is directly related to communication range and bit error rate (BER). It is defined as the signal level required for a particular *quality* of received information to be achieved. In the case of digital radios, the parameter defining quality is the bit error rate, at a specific signal-to-noise ratio. Sensitivity is the absolute input power level that gives the desired signal-to-noise ratio. Its computation is based on the *minimum detectable signal* (MDS) and required carrier-to-noise ratio [1]. When referenced to the input, the sensitivity is expressed as the sum of the MDS and the required output signal-to-noise ratio,

$$Sensitivity(dBm) = \text{MDS}(dBm) + \frac{C}{N} \tag{6.1}$$

where, *C/N* is the output signal-to-noise ratio that produces the desired performance [1].

6.2.1 *Minimum Detectable Signal*

The minimum detectable signal may be defined with respect to the noise floor at the IF, Fig. 6.3. The noise floor at the IF is obtained as follows [1]. Assuming a temperature T_0 at the receiver input, Fig. 6.3, and a corresponding input noise level of,

$$N_{in} = kT_0B \tag{6.2}$$

if the receiver possesses a gain G_{RX}, a noise power, N_{RX}, a noise factor F_{RX}, and a bandwidth B, then the output noise is,

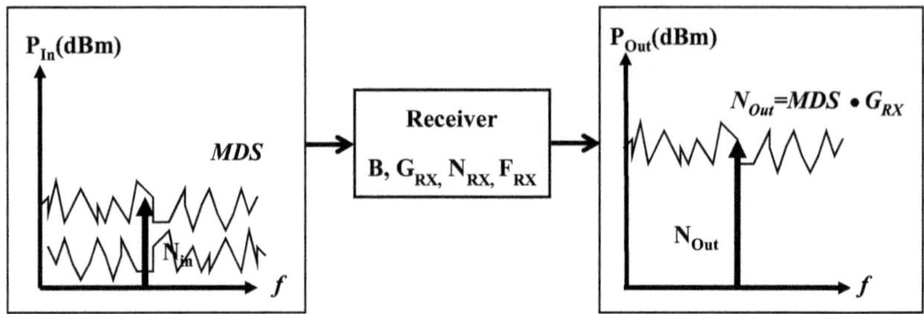

Fig. 6.4 Illustration of minimum detectable signal (MDS). (*After* [1])

$$N_{Out} = (N_{RX} + kT_0 B)G_{RX} = F_{RX} kT_0 B G_{RX} \qquad (6.3)$$

which expresses the fact that the receiver takes the input noise power, adds to it the receiver noise power, and amplifies both together by the receiver gain to produce the output noise. Now, with respect to the input noise, the receiver noise power may be expressed by,

$$N_{RX} = (F_{RX} - 1)kT_0 B \qquad (6.4)$$

so that,

$$N_{Out} = F_{RX} kT_0 B G_{RX} = F_{RX} \cdot kT_0 (1Hz) \cdot \left(\frac{B}{1Hz}\right) \cdot G_{RX} \qquad (6.5)$$

which, expressed in decibels, takes the form,

$$10\log\left(N_{Out}\right) = 10\log\left[kT_0 (1Hz) \cdot \left(\frac{B}{1Hz}\right) \cdot F_{RX} \cdot G_{RX} \right]$$

$$= -174dBm + 10\log B + N_{RX} + G_{RX} \, (dB) \qquad (6.6)$$

This equation is equal to the signal power N_{Out} delivered to the output, i.e., it is equal to the *output noise floor*. Thus, the MDS is defined as, Fig. 6.4,

$$MDS = -174dBm + 10\log B + N_{RX} \qquad (6.7)$$

Again, the noise at the input passes through the receiver bandwidth, gets amplified by the receiver gain and, upon addition to the receiver's own noise, appears at the output as noise floor, N_{Out}. If, from the output noise power in dBm, we subtract the receiver gain in dB, we obtain the input MDS. Thus, the MDS equation determines the input signal level required to deliver an output signal to a load equivalent to the output noise floor, Fig. 6.4 [1].

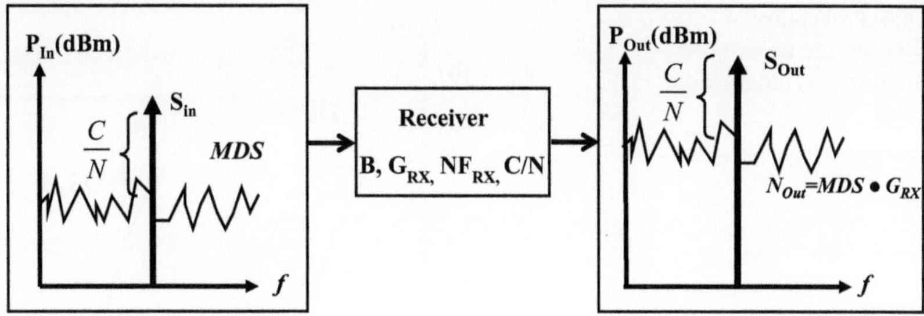

Fig. 6.5 Illustration of SNR/CNR. *After* [1]

6.2.2 Signal-to-Noise Ratio

The MDS provides the output signal power that equals the output noise floor. Most receivers, however, must output a signal that is *above* the noise floor, from which the information can be extracted. This means that a specific signal-to-noise ratio is necessary in order to recover the received information with a specific level of quality. The necessary signal-to-noise ratio is also referred to as *carrier-to-noise ratio*, C/N. It must be clarified that, while MDS references the source noise and noise added by the receiver to its input terminals, this is not the actual input noise floor but, rather, an imaginary noise floor due to noise added by the receiver (referred to the input). The real noise floor is actually the source noise given by kT_0B when the receiver is terminated at its input by a passive network.

As shown in Fig. 6.5, when the MDS is used as the noise reference at the input, the carrier-to-noise level is the same at both the input and the output. In this context, the input signal S_{in} must be above the MDS level by a certain amount, which means that the output signal S_{Out} is above the output noise floor also by this magnitude. Therefore, since the receiver adds noise, when the source noise is utilized as reference, the real signal-to-noise ratio at the input is greater than the signal-to-noise ratio at the output [1].

The signal-to-noise ratio required depends on several system parameters that contribute to the signal-to-noise ratio. These include, the modulation scheme, the distortion of the group delay in the IF filters, detector linearity, and distortion [1].

The desired quality of the information received also plays a large role in determining the carrier-to-noise required. For example, in a digital receiver, the bit-error rate (BER), or the probability P(e) that any received bit is in error, is the primary measure of quality. BER is defined as the number of bit errors divided by the number of bits transmitted, and is degraded by the jitter, which results from noise in the receiver. Other contributors to BER include the detector linearity, the frequency response of the filter, and the bandwidth. The BER has a nonlinear relationship to the receiver's MDS, Fig. 6.6. In particular, the greater the amount by which the desired signal exceeds the noise floor, the lower the BER (P(e)). It is found, then, that the probability of error increases abruptly as signal-to-noise ratio decreases, which results in BER changes by several orders of magnitude with small increments in the

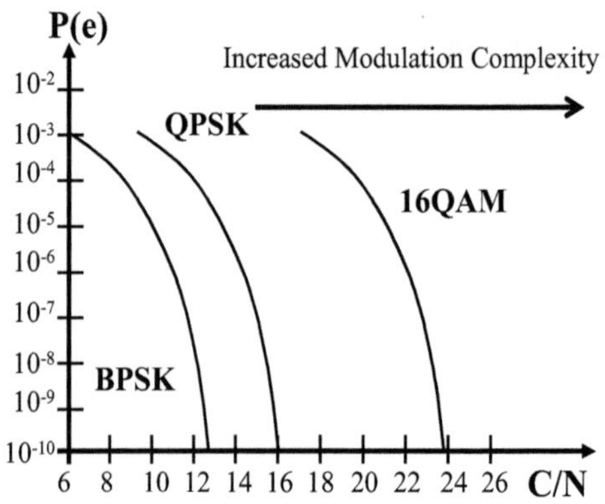

Fig. 6.6 Probability of error versus carrier-to-noise ratio for various modulation schemes

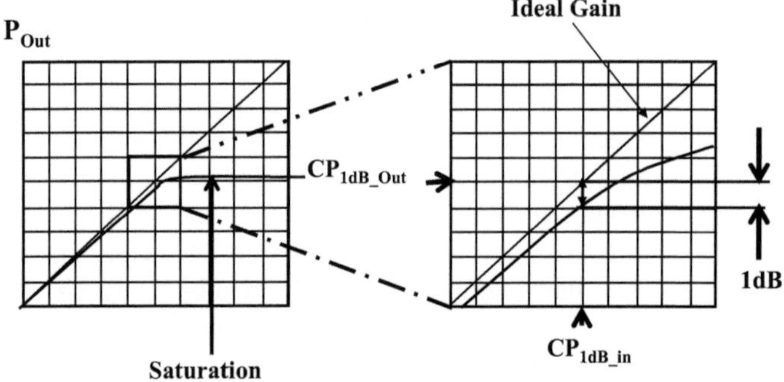

Fig. 6.7 General output power versus input power characteristic

signal-to-noise ratio. On the other hand, the BER worsens with a decreased signal level. In general, as shown in Fig. 6.6, to achieve an acceptable bit error rate, modulation schemes with increased complexity require higher signal-to-noise ratios [1].

6.2.3   1-dB Compression Point

As the amplitude of the input signal entering a practical two-port network increases, the corresponding output signal increases linearly until the energy from distortion products substantially combines with the fundamental output power. The result is that distortion products, along with fundamental, appear at the output, i.e., the strictly linear output power change versus an input power change is no longer observed, Fig. 6.7.

Thus, the power gain given by the ratio $G = P_{Out}/P_{in}$ is no longer constant, Fig. 6.8.

To characterize the deviation from linearity in the gain versus power plot, the concept of *1-dB compression point* ($P_{1\text{-dB}}$ CP) is used. This is the input signal level at which the corresponding output power is 1 dB below what it would have been in

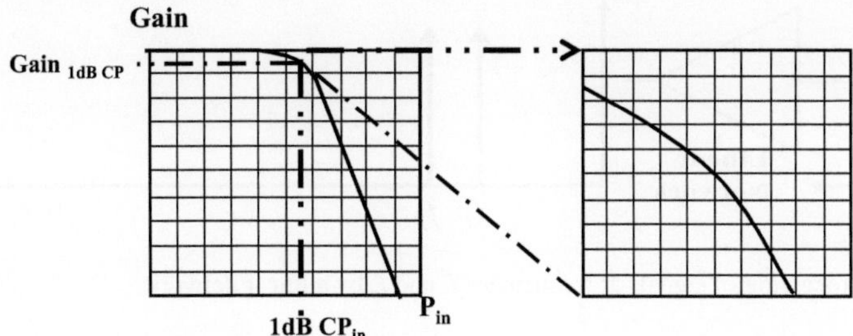

Fig. 6.8 Gain versus input power

a strictly linear circuit. As the input signal's amplitude continues to increase, so does the number of higher-order harmonics contributing to the output. In particular, the output power versus input power continues to deviate from the ideal linearity until, eventually, the input signal increases to a point where the output power corresponding to the fundamental frequency reaches a maximum value; this point is the *saturation point*. Beyond this point, further input signal increases cause the gain to drop (or the loss to increase), Fig. 6.8. Once the circuit it "compressed," gain decreases for all signals being processed through it. A strong signal input (either desired or undesired) into a receiver can, therefore, cause compression.

The importance of compression may be gauged from the following. If a weak desired signal is input into a receiver, along with a strong undesired signal, it is observed that the strong undesired signal causes compression. Then, if the desired signal is at the sensitivity level, a decrease in gain reduces the carrier-to-noise level which, in turn, increases the bit error rate, Fig. 6.6.

The 1-dB gain compression point is defined as the amount of power required to cause a gain drop of 1 dB due to circuit nonlinearity, and it may in general be referenced either to the input or output. When referenced to the input, (CP_{1dBin}) indicates the amount of power incident at the input of the system that reduces the small signal gain of equation $G(dB) = 10\log(P_{Out}/P_{in})$ by 1 dB. The relationship between the output and input 1-dB compression points, CP_{1dBou} and CP_{1dBin}, is given as follows [1],

$$CP_{1dBout} = CP_{1dBin} + Gain - 1dB \qquad (6.8)$$

or

$$CP_{1dBin} = CP_{1dBout} - Gain + 1dB \qquad (6.9)$$

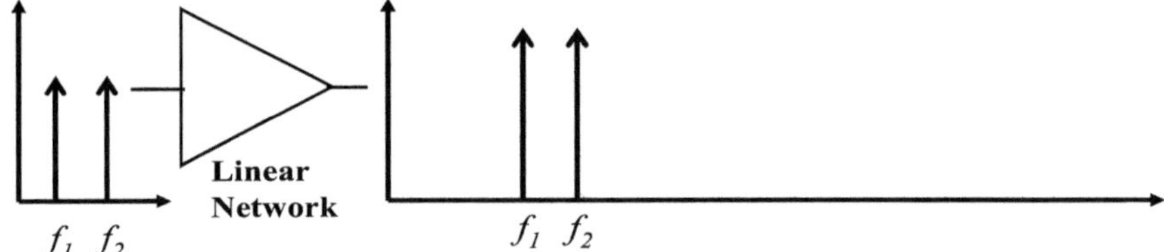

Fig. 6.9 Processing of signals at frequencies f_1 and f_2 by a linear network

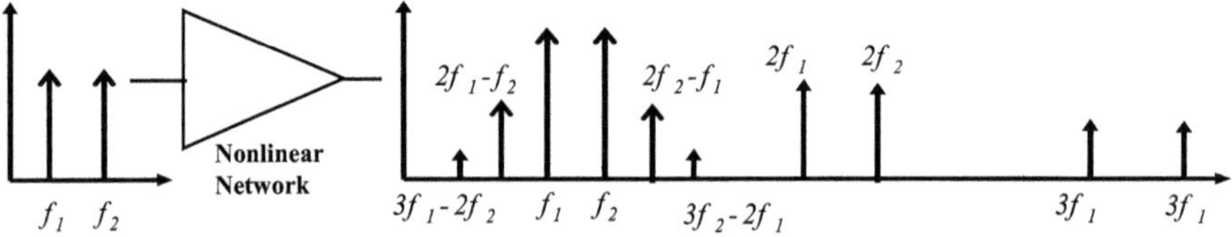

Fig. 6.10 Processing of signals at frequencies f_1 and f_2 by a nonlinear network

6.2.4   Intermodulation Distortion

In addition to compression, nonlinearity in receiver systems gives rise to *distortion products*; these are frequency components that are easily identifiable in the context of the signal spectrum at the output of a linear system. In particular, two RF signals, f_1 and f_2, applied at the input of a linear amplifier, Fig. 6.9, result in both signals being amplified and delivered to the output load.

Since the system is assumed linear, changes in the power of both input signals yields a corresponding output power change. That is, increasing the input power by 1 dB results in a corresponding increase of 1 dB in the output power of the fundamental. On the other hand, if the two RF signals, f_1 and f_2, are applied to a nonlinear amplifier, it is found that additional signals, known as *distortion products*, are generated internally and delivered to the load, Fig. 6.10. These output products include a sum term $f_1 + f_2$, a difference term $f_2 - f_1$, the second harmonics $2f_1$ and $2f_2$, etc. Additionally, mixing products are generated at predetermined specific frequencies given by,

$$IMD = \pm m f_1 \mp n f_2 \tag{6.10}$$

From (6.10), the *third-order IMD* products (i.e., those for which $m + n = 3$) occur at frequency locations $2f_2 - f_1$ and $2f_1 - f_2$. Thus, the second harmonic of f_2 interacts with fundamental f_1 and the second harmonic of f_1 interacts with fundamental f_2. As seen in Fig. 6.10, these distortion products are spaced equally with respect to the two fundamental frequency terms, and exhibit the behavior that, when the power levels of f_1 and f_2 are changed by *1 dB*, that of both third-order IMD products change by

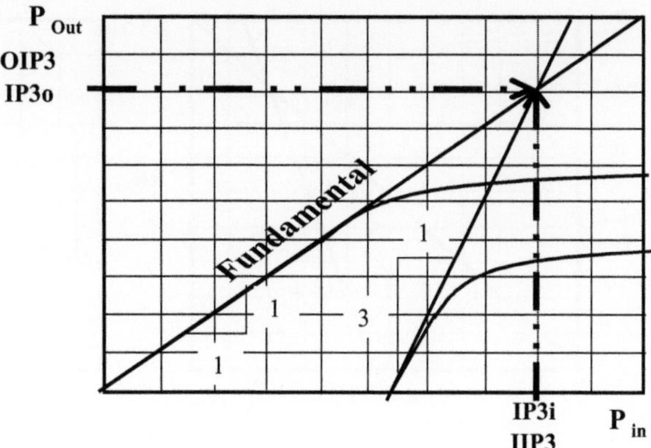

Fig. 6.11 Output power Pout for one fundamental frequency signal versus input power P_{in}

3 dB, Fig. 6.11. Clearly, the amplitude of the third-order products is lower than that of the fundamental terms. This difference, while dependent on the input power, is named *rejection ratio* (RR) [1]. It is observed that, as the input power is increased, for every 1 dB of increase, the third-order intermodulation rejection ratio decreases by 2 dB. Ideally, the continued increase in input power level eventually results in a distortion product power that equals that of the fundamental frequency. This power is called the *third-order intercept point* (IP3).

A plot of the output power at the fundamental versus the input power has a linear characteristic with unity slope as shown, Fig. 6.11. As seen, the fundamental output power versus input power has a linear characteristic with unity slope.

Figure 6.11 depicts a plot of a third-order intermodulation distortion product together with the fundamental, showing that the third-order intermodulation distortion product has a linear characteristic with a slope of three. Because the slope of the distortion product is steeper than that of the fundamental, the third-order distortion characteristic eventually intersects the fundamental curve. This intersection is known as the network's *third-order intercept point*. The intercept point may be referenced at the network's input or output. When referenced to the input it is denoted IP3i or IIP3, whereas, when referenced to the output is denoted as IP3o or OIP3 [1].

A number of relationships exist among the various parameters characterizing distortion. For instance, the output intercept point is equivalent to the input intercept point plus the small-signal gain in dB as expressed by,

$$OIP3 = IIP3 + Gain \qquad (6.11)$$

It should be kept in mind that the intercept concept is an ideal one in the sense that it is impossible for the input power of a real system to be large enough to reach the theoretical intercept point; compression would set in before that happens. Upon the beginning of compression, the resulting increase in power of the fundamental frequency and the distortion products is limited with further increases in input power. In practice, it is found that the 1 dB compression point is usually around 10 dB

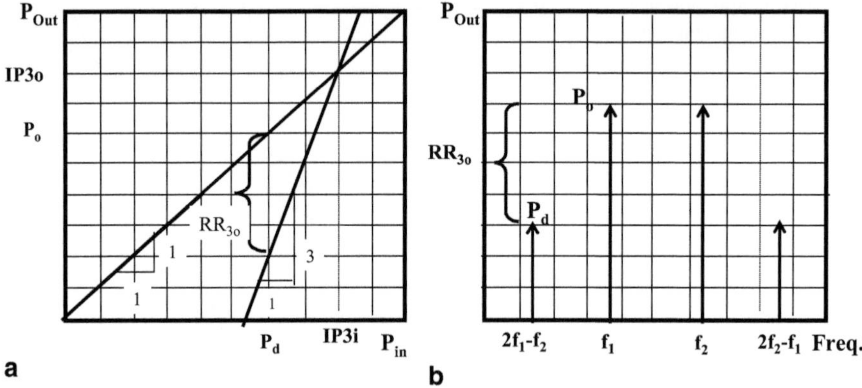

Fig. 6.12 a Fundamental P_o and third-order distortion P_d output powers versus input power P_{in}. **b** Output frequency spectrum at a fixed input power level

below the third-order intercept point. Therefore, intercept point measurements must be performed at power levels at which the device is not compressed [1].

Another distortion parameter, the third-order rejection ratio (RR_{3o}), is measured using a spectrum analyzer at a fixed input power level, Fig. 6.12 [1].

As depicted in Fig. 6.12, RR_{3o} is the difference in dB between the output powers at the fundamental P_o and at the distortion P_d, taken at one point on the third-order distortion characteristic. The RR_{3o}, referenced to the output, is defined on the output power versus input power plot. The rejection ratio RR_{3o} for a given power level is needed to determine whether undesired signals in a receiver environment will interfere with it.

The output third-order intercept point, on the other hand, is obtained by measuring the output power and rejection ratio [1–3],

$$IP_{3o} = P_o + \frac{RR_{3o}}{2} \tag{6.12}$$

or

$$OIP3 = P_o + \frac{RR_{3o}}{2} \tag{6.13}$$

The third-order rejection ratio RR3o is simply the dB difference between the measured output power of one of the fundamental tones and one of the distortion products,

$$RR_{3o} = P_o - P_d \tag{6.14}$$

Substituting (6.14) into (6.12) yields the third-order output intercept point of the two-port,

$$OIP3 = P_o + \frac{RR_{3o}}{2} = P_o + \frac{P_o - P_d}{2} = 1.5P_o - 0.5P_d \qquad (6.15)$$

Therefore, measuring the fundamental output power P_o and the distortion power P_d determines the third-order intercept point referenced to the output.

Similarly, by rearranging the equation for OIP3, the third-order input intercept point IIP3 is determined,

$$IIP3 = OIP3 - Gain \qquad (6.16)$$

Once the third-order output intercept point is known, the distortion product power level is determined for other input power levels as follows,

$$OIP3 = 1.5P_o - 0.5P_d \rightarrow P_d = 3P_o - 2OIP3 \qquad (6.17)$$

All this effort, spent discussing third-order products, may make the reader wonder: Why is the third-order intercept point an important performance parameter? There are a number of reasons for this, related to the environment in which the receiver operates. Firstly, in principle, all RF signals whose frequencies happen to coincide with those of the selected receiver channels, will enter and be processed by the receiver. Secondly, in addition to these there will exist, in general, adjacent and close-in signals that will be processed through the nonlinear circuits in the receiver's front end, such as the amplifier and mixer circuits. Thirdly, because signals in close channels enter and are processed, they create third-order distortion products *within* the passband coincident with the RF front end. Fourthly, these distortion products may occur at the *same frequency as the desired signal.* And, therefore, fifthly, if these distortion products possess a power level slightly less than, or at a power anywhere above the desired signal, they will interfere with its reception! This may be surmised from an examination of the spectrum in Fig. 6.13a, corresponding to an assumed perfectly linear receiver. Here, the spectrum shows the desired signal f_d and two undesired signals f_{U1} and f_{U2}, with the closest undesired signal f_{U1} assumed to be located one channel spacing away from the desired signal, say, 30 kHz, and the second undesired signal f_{U2} located two channel spacings away from the desired signal, say, 60 kHz. In this case, because of the assumed perfectly linear receiver, interference from third-order distortion products is not possible.

On the other hand, when the receiver under consideration is nonlinear, its output spectrum would look like that in Fig. 6.13b. In this case, the two undesired signals f_{U1} and f_{U2} cause a third-order intermodulation distortion product that is *coincident* with the desired signal such that, if the distortion product power is greater than the desired signal power, the reception of the desired information at f_d is precluded. Under this circumstance, due the coincidence of desired and third-order product frequencies, even filtering cannot ensure clear reception. Therefore, the intercept point must be high enough to provide *a third-order rejection ratio* capable of protecting or guaranteeing that the largest interfering signals incident at the receiver input terminals would not give rise to products exceeding the power of the desired signal [1–3].

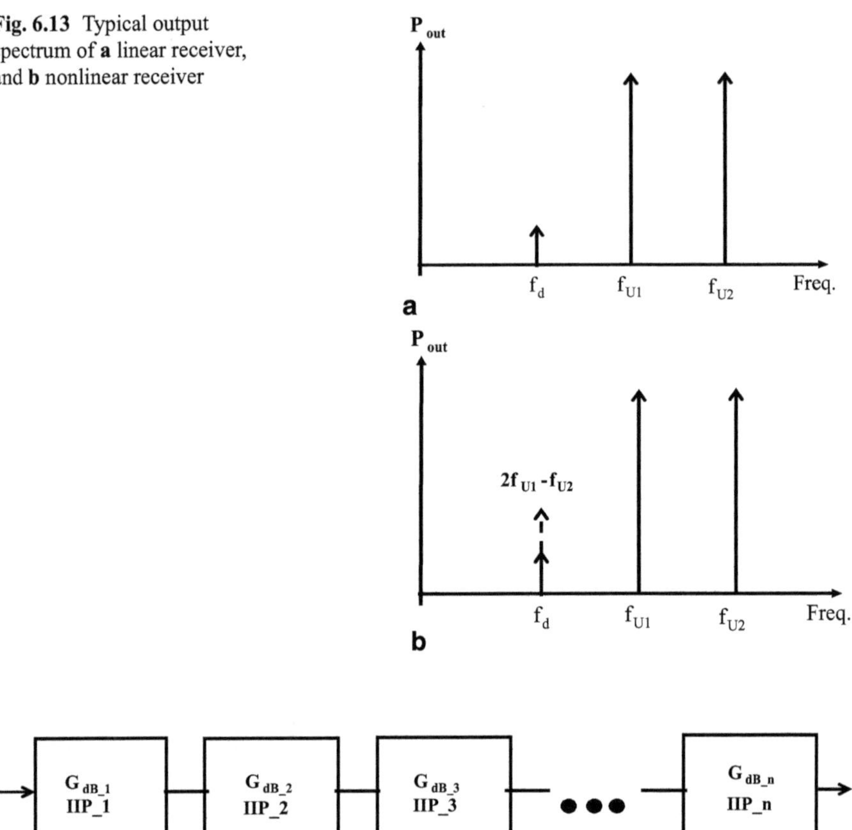

Fig. 6.13 Typical output spectrum of **a** linear receiver, and **b** nonlinear receiver

Fig. 6.14 Cascaded receiver networks

6.2.5 *Cascaded Receiver Networks*

In this sub-section we address the calculation of the third-order intercept point for a receiver system assumed to consist of a number of cascaded building blocks, each one characterized by gains and third-order input intercept points, Fig. 6.14. The calculation may be utilized for analysis, when the gains and intercept point are known, or for design, when trials are undertaken to ascertain the parameters that will produce an overall desired performance target.

In particular, it will be seen that each building block in the system manifests its presence by impacting the total system distortion, and that this may occur even when it adds no distortion of its own. The total *numeric* intercept point IIP_T for the cascaded system may be shown to be given by the reciprocal of equation,

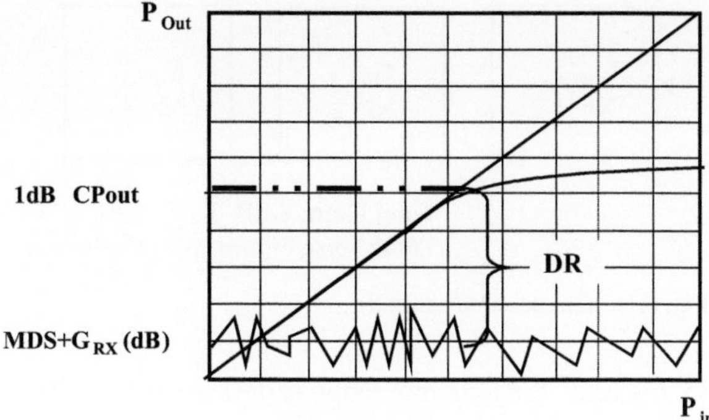

Fig. 6.15 Dynamic range (DR) definition

$$\frac{1}{IIP\_T} = \left[\left(\frac{1}{IIP\_1} \right)^q + \left(\frac{g_1}{IIP\_2} \right)^q + \left(\frac{g_1 g_2}{IIP\_3} \right)^q + ... \left(\frac{g_1 g_2 \cdots g_{n-1}}{IIP\_n} \right)^q \right] \quad (6.18)$$

where $q = \dfrac{m-1}{2}$ and m is the slope of the response ($m = 3$ for third order). In effecting calculations with this equation, it must be realized that the gains and input intercept points (g_i's and IIP_i's) refer to numerical entities, rather than dB of dBm [1, 2].

6.3   Dynamic Range

The dynamic range is calculated by subtracting the MDS from the 1-dB compression point referenced to the input, Fig. 6.15, given by,

$$DR = CP_{1dBin} - MDS_{dBm} = CP_{1dBin} + 174 \ dBm - 10\log(B) - NF_{RX} \quad (6.19)$$

In this equation, the 1-dB compression point CP1dBin is given in dBm. It is often found that the receiver dynamic range is modified from the above definition, which uses MDS, on the low end, to a definition using sensitivity, on the low end. This alternative definition gives a value less than the definition of the previous equation by an amount equal to the required carrier-to-noise ratio [1],

$$DR = CP_{1dBin} - Sensitivity = CP_{1dBin} + 174 \ dBm - 10\log(B) - NF_{RX} - \frac{C}{N} \quad (6.20)$$

The top end of the dynamic range definition may also be modified for receiver systems. Instead of using the 1-dB compression point, what is used is the highest power that the receiver can handle and still meet a particular level of performance.

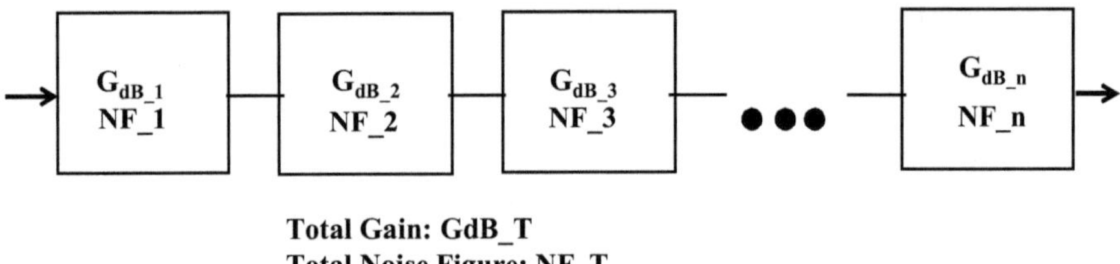

Fig. 6.16 Cascaded noise figure

In this case, the dynamic range is defined as the difference between this power and the sensitivity power level [1, 2].

The dynamic range may also be defined as the difference between the largest signal that the receiver can reject while simultaneously receiving a smaller signal at the sensitivity level. The effect of this largest signal, i.e., the loss of the receiver's ability to detect a signal at the sensitivity level, is called *receiver blocking* [1, 2] and is addressed next.

6.3.1 Receiver Blocking

Large undesired signals within the receiver's RF passband cause gain compression of certain stages. This gain compression, in turn, reduces the signal power delivered to the load for all signals, including the desired signal. Now, if the system gain drops, then the total system noise figure may increase, depending on which stage in the cascade sequence is compressed. Under this circumstance, the equation (6.21) below may be utilized to predict the system noise figure as a function of distributed gains, Fig. 6.16.

$$F_T = F_1 + \frac{F_2 - 1}{G_1} + \frac{F_3 - 1}{G_1 G_2} + ... + \frac{F_n - 1}{G_1 G_2 \cdots G_{n-1}} \tag{6.21}$$

In general, if the noise figure degrades, the MDS and sensitivity levels degrade as well. However, it may be the case that, depending on the stage in question, the overall noise figure of the system may not degrade significantly; in any event, if gain compression is experienced, the amount of desired signal delivered to the load is reduced. So, both effects degrade signal-to-noise ratio.

Consider now a receiver with a desired signal at the receiver's sensitivity level, Fig. 6.17.

As mentioned above, when a large undesired signal is introduced within the RF passband, then the receiver gain drops due to compression. This gain drop, in turn, causes a reduction in the desired signal, thereby lowering the signal-to-noise ratio. In addition, this reduction causes the desired signal to drop below the sensitivity level. Since the sensitivity level requires a particular signal-to-noise ratio for the information in the received signal to be recoverable at the desired performance level, the reduction in signal-to-noise ratio thus causes a loss in the amount of information

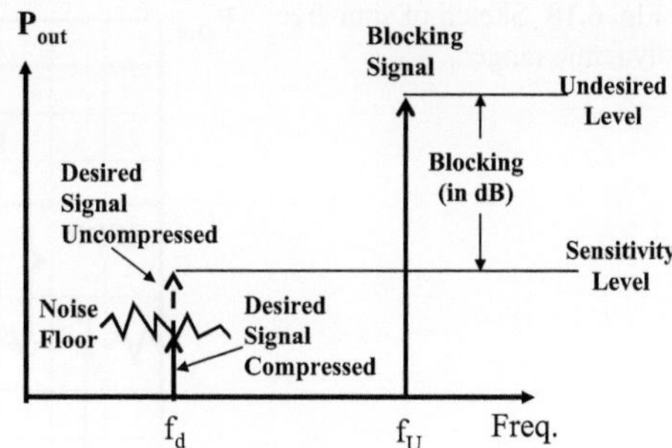

Fig. 6.17 Receiver blocking

extracted. Blocking is defined as the amount of power required to cause the degradation of receiver sensitivity from compression effects. It is usually characterized with an undesired signal located 1 MHz away from the desired signal. In particular, the difference between the power of the undesired signal and the sensitivity level captures the parameter denoted *blocking* and is expressed in dB, Fig. 6.17 [1].

6.3.2 Spur-Free Dynamic Range

The so-called Spur-free dynamic range (SFDR) parameter is similar to the third-order intermodulation distortion rejection ratio. In particular, the third-order intermodulation rejection ratio is defined as the power difference between the power at the fundamental and that at the distortion product, which is given by equation,

$$RR_{3o} = P_o - P_d \tag{6.22}$$

where, the rejection ratio depends on a specific output power, P_o. A change in fundamental power, therefore, produces a change in rejection ratio. As shown previously, a 1-dB change in fundamental power is accompanied by a 3-dB change in the power of the third-order distortion product and, since the fundamental has a slope of unity and the distortion product has a slope of 3, the rejection ratio exhibits a slope of 2 versus output power of the fundamental. Therefore, a 1-dB change in the power of the fundamental changes the rejection ratio by 2 dB. The spur-free dynamic range is defined as the difference between the power of the fundamental and the noise power when the distortion products are equal to the noise power, Fig. 6.18.

The spur-free dynamic range (SFDR) is expressed by equation [1–3],

$$SFDR = \frac{2}{3}[OIP3 - MDS - G_{RX}] = \frac{2}{3}[OIP3 + 174dB - 10\log(B) - NF_{RX} - G_{RX}] \tag{6.23}$$

We present next an example of trade-offs encountered in a receiver design as a result of architectural decisions, Fig. 6.19. In particular, this example compares the sensitivity and the spur-free dynamic range when a RF filter follows the RF amplifier and when the RF filter precedes the RF amplifier.

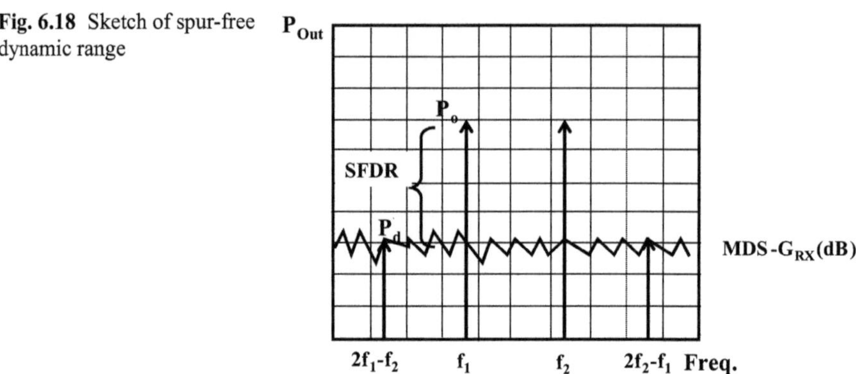

Fig. 6.18 Sketch of spur-free dynamic range

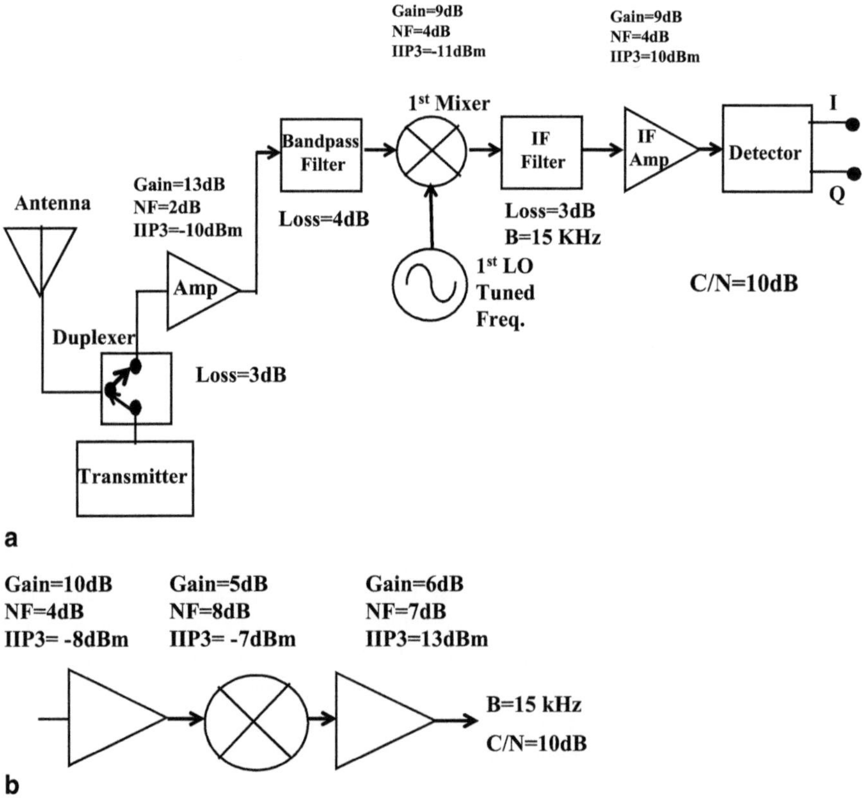

Fig. 6.19 a Receiver block diagram. b Six-stage network combined into three-stage network

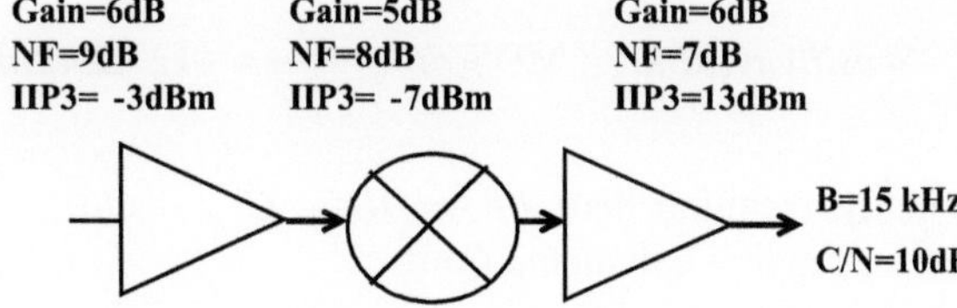

Fig. 6.20 Six-stage network combined into three-stage network when the filter is moved in front of the amplifier

First, the total gain for the receiver G_{RX} is calculated by simply adding together the gains of all the stages,

$$G_{RX} = -3+13-4+9-3+9 = 21\,dB \tag{6.24}$$

Next, the losses in front of the gain stages are combined with those of the gain stages, as shown in Fig. 6.20; doing this to the six-stage cascade, Fig. 6.20a, combines into a three-stage network, Fig. 6.20b. In this combination, loss in front of a gain stage decreases the gain dB for dB and increases the noise figure and the input intercept point dB for dB. This is true for the noise figure as long as additional noise is not impressed from external sources into the lossy stage, and for the intercept point if the loss does not generate distortion products of its own (that is, it is linear). Next all dB values of gain, noise figure, and input intercept point are converted to numerical values. The system numerical gains are given by $g_1 = 10$; $g_2 = 3.16$; $g_3 = 4$, and the noise factor of each stage is: $f_1 = 2.5$, $f_2 = 6.3$, $f_3 = 5$. The numerical input intercept points of each stage are given by: IIP3_1 = 0.158, IIP3_2 = 0.2, IIP3_3 = 19.95. Next, the noise factor of the receiver is given by,

$$F_T = 2.5 + \frac{4-1}{10} + \frac{6.3-1}{10(3.16)} = 3.12 \tag{6.25}$$

Converting this to noise figure yields NF = 4.95 dB. The third-order input intercept point is calculated using,

$$\frac{1}{IIP\_T} = \left[\left(\frac{1}{0.158}\right)^1 + \left(\frac{10}{0.2}\right)^1 + \left(\frac{10 \cdot 3.16}{19.95}\right)^1\right] = 57.913 \tag{6.26}$$

so that,

$$IIP\_T = 10\log(0.017) = -17..7\,dBm \tag{6.27}$$

The corresponding sensitivity is calculated from,

$$MDS = -174\,dBm + 10\log B + N_{RX}$$
$$= -174\,dBm + 10\log(15000)\,dB + 4.95\,dB = -127.29\,dBm \tag{6.28}$$

$$Sensitivity(dBm) = \text{MDS}(dBm) + \frac{C}{N} = -127.29dBm + 10 = -117.29dBm \quad (6.29)$$

Finally, recalling that, $IIP3 = OIP3 - G_{RX} \rightarrow OIP3 = IIP3 + G_{RX}$, the spur-free dynamic range is calculated from,

$$SFDR = \frac{2}{3}[OIP3 - MDS - G_{RX}] = \frac{2}{3}(-17.7 + 127.289) = 73.06dB \quad (6.30)$$

How is the above performance altered if the filter is moved in front of the amplifier, so that the new combined receiver is as shown in Fig. 6.20?

After carrying out the calculations above, the new sensitivity and spur-free dynamic range are, −112.38 dBm and 65.87 dB, respectively, which shows that both are degraded by moving the RF filter in front of the amplifier. This is an example of the type of analysis one engages in during radio system design.

6.4   Summary

In this chapter the topics of sensitivity and dynamic range for a receiver were addressed. We began by introducing their definitions, the concept of minimum detectable signal, signal-to-noise ratio, 1 dB compression point (CP 1 dB), saturation, and some relationships between input and output CP 1 dB), intermodulation distortion, third-order intercept point (IP3), receiver blocking, and spur-free dynamic range. The relationships between these parameters as applied to building blocks in a cascade arrangement making up a receiver, and the overall parameters for the receiver, were presented. The chapter concluded with an example of how changes in the sequence of building blocks in a cascade impact the sensitivity and dynamic range of a receiver.

References

1. *Texas Instrument Technical Brief SWRA030*, Matt Loy, Editor, May 1999.
2. Q. Gu, *RF System Design of Transceivers for Wireless Communications*, Springer, 2005.
3. R. Sagers, Intercept Point and Undesired Responses, IEEE Trans. on Vehicular Tech., Vol. VT-32, No. 1, Feb. 1983, pp. 121–133.

Chapter 7
Transceiver Architectures

Abstract This chapter addresses the topics of transmitter and receiver architectures, and the performance of oscillators. In particular, the heterodyne and homodyne architectures are discussed, together with the image-reject, digital-IF, and sub-sampling architectures. The phase noise performance of oscillators is then discussed, followed by the impacts of oscillator pulling and pushing. The chapter concludes with a pictorial presentation of the manifestations of phase noise on the spectrum of a down converter, and of timing jitter on several digital modulation schemes, in particular, BPSK, QPSK and 16QAM.

7.1 Introduction

The choice of a transmitter architecture is determined by two important factors, namely, wanted and unwanted emission requirements, and the number of oscillators and external filters. In general, the architecture and frequency planning of the transmitter must be selected in conjunction with those of the receiver so as to allow sharing hardware and possibly power. Thus, in modern wireless communications it is common to talk about transmitters-receivers, or *transceiver*, Fig. 7.1.

There are two dominant transceiver architectures: *heterodyne* and *homodyne*. The heterodyne (**Hetero: different; Dyne: to mix**) architecture is the most popular, and is based on the process of mixing an incoming RF signal with an offset-frequency local oscillator (LO) in a nonlinear device to generate an intermediate frequency (IF) signal in the receiver or to produce an RF signal from its IF version in the transmitter [1, 2, 3, 4, 5, 6, 8]. The nonlinear device in question is called a frequency mixer or frequency converter. In general, the frequency translation processes may be performed more than once, thus, there may exist multiple intermediate frequencies and multiple IF blocks.

In the homodyne (**Homo: same; Dyne: to mix**) architecture, frequently called Zero-IF/Direct Conversion, an RF signal is directly down-converted to a baseband (BB) signal, or vice versa, without any intermediate frequency stages and, therefore, it is also referred to as zero IF architecture. The direct-conversion receiver has

H. J. De Los Santos et al., *Radio Systems Engineering,*
DOI 10.1007/978-3-319-07326-2_7, © The Authors 2015

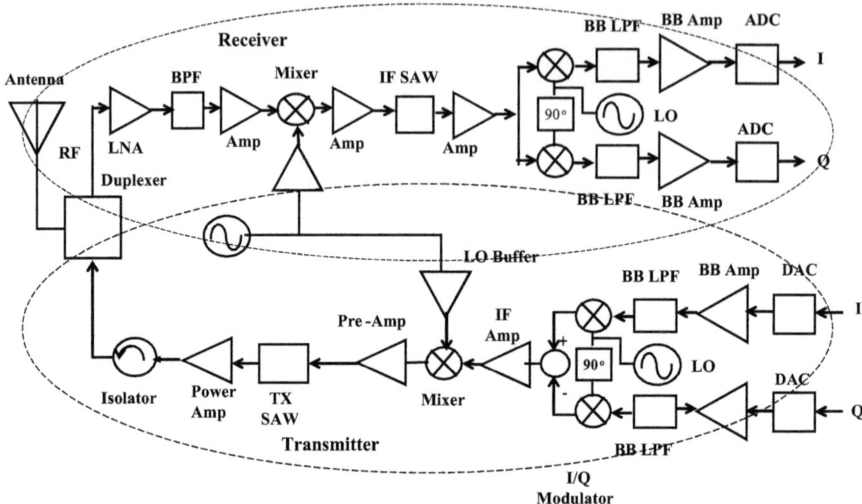

Fig. 7.1 Block diagram of super-heterodyne full-duplex transceiver

many attractive features. In the first place, it has no IF and, thus, the expensive IF passive surface acoustic wave (SAW) filter can be eliminated; as a result, the cost and size of the overall transceiver are reduced. In the second place, the channel filtering is implemented in the analog BB by means of a low-pass filter. Therefore, its bandwidth (BW) is adjustable and it is easy to design the direct-conversion receiver for multimode operation with a common analog baseband circuitry and even a common RF front-end from the pre-selector to the RF down-converter if all the modes operate in the same frequency band.

The configuration of the direct-conversion radio looks simpler than that of the heterodyne, but its implementation is much more difficult [4, 5]. In the next discussion we will explore this.

7.2   Transmitter Architectures: Heterodyne

A block diagram of a heterodyne transmitter is shown in Fig. 7.2 [4]. The diagram is conceptually divided into three sections: baseband (BB) intermediate frequency (IF), and radio frequency (RF). The elements of this architecture are explained next.

Beginning on the right-hand side of Fig. 7.2, and stepping through towards the left-hand side, we have the BB, IF, and RF sections. In the baseband section, in-phase (I) and quadrature (Q) digital representations of the information signal to be transmitted are converted to analog signals by digital-to-analog converters (DACs).

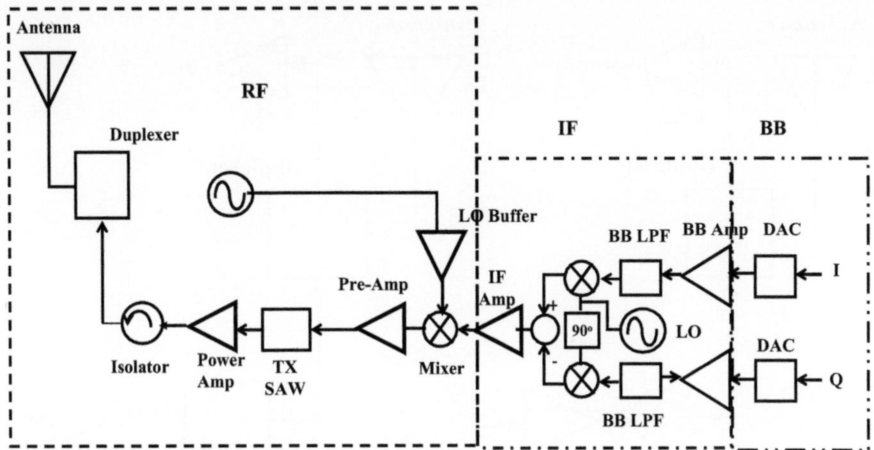

Fig. 7.2 Block diagram of super-heterodyne transmitter

After BB filtering, the I and Q BB signals are up-converted into IF signals. In the I/Q modulator, during the frequency up-conversion, the Q channel IF signal is phase-shifted by an additional 90° with respect to the I channel. The composite IF signal is then amplified by amplifier IF Amp (usually a variable gain amplifier-VGA), which typically consists of multiple stages.

An up-converter follows the IF VGA and the amplified IF signal is then translated up to an RF signal. The RF signal is further amplified by an RF VGA, and then by a driver amplifier, to a power level high enough to drive the power amplifier (PA). An RF BPF (SAW filter) is inserted between the driver and the PA to select the desired RF signal and suppress other mixing products generated by the RF up-converter. One would think that it would be better to place the RF BPF immediately after the up-converter, but this arrangement would not be convenient since the whole block, from the ADC to the IF Amp driving the mixer, may be integrated on a single semiconductor chip.

In the RF section, depending on the wireless standard, a Class AB or Class C PA is employed [5]. The Class C PA is capable of higher power efficiency than the Class AB PA, but it can only be used for constant envelope modulation schemes such as FM and GMSK [5].

The PA performance, in particular, its gain and nonlinearity, are very sensitive to its load and, therefore, an isolator is often used between the PA and the antenna to reduce the influence of variations in the antenna environment [4, 5], which affect its input impedance and, thus the PA load, Fig. 7.3a. Modern concepts under evaluation exclude the isolator, but replace it with a tunable matching network, Fig. 7.3b.

In Fig. 7.3, a duplexer filter separates the transmit (TX) and receive (RX) bands, in the case of frequency-domain duplexing (FDD). If the TX and RX bands coincide, the duplexer is replaced by an RF switch to perform time-domain duplexing (TDD). The duplexer filter suffers from a loss of several dB, whereas the switch

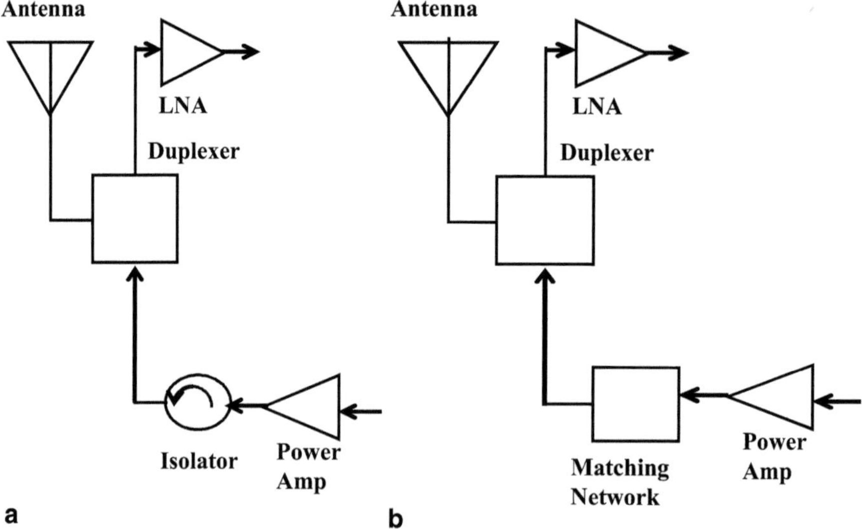

Fig. 7.3 PA/Antenna interface. **a** Using duplexer-PA isolator. **b** Using duplexer-PA matching network

introduces a typical loss of the order of 1 dB [5]. In any event, the loss in these components reflects on a lower effective PA efficiency, since additional bias current would be necessary to compensate for it [5]. The LO synthesizer of the RF section provides the LO power to the RF up-converter, and effects channel tuning.

7.3 Transmitter Architectures: Homodyne

A typical block diagram of the homodyne transmitter architecture is shown in Fig. 7.4.

In the baseband, the I and Q BB signals coming from the DACs are passed through low-pass filters to make adjacent channel and alternate-channel emission levels further suppressed, and to eliminate aliasing products [5]. The filtered I and Q BB signals are both directly up-converted to RF, and then added by an I/Q modulator [5].

In the RF section, the homodyne, direct conversion, or zero-IF (no intermediate frequency) transmitter translates the desired baseband frequency directly to RF for transmission frequency. The composite RF signal is amplified all the way up to the RF PA. This is followed by a BPF which is inserted between the driver amplifier and the PA to suppress the out-of-band signals, especially those in the receiver band, noise and spurs emissions [4, 5]. Compared to the super-heterodyne transmitter, the homodyne does not save any passive filters, except the LO synthesizer. The outstanding feature of the homodyne transmitter is that its transmission contains

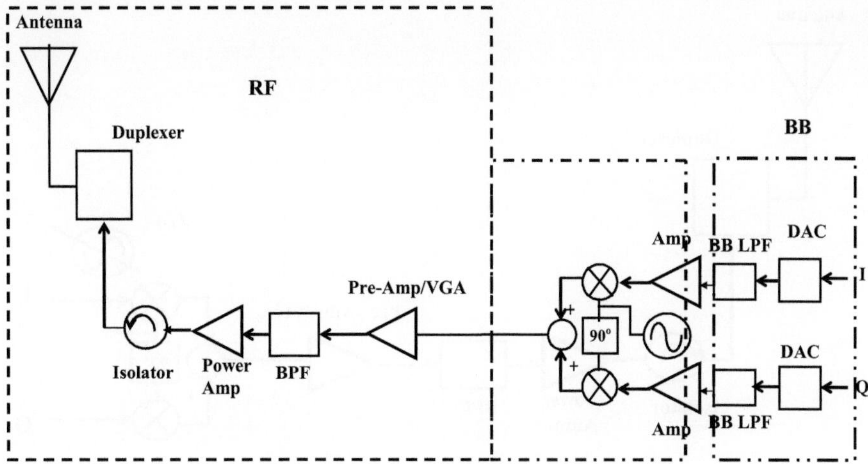

Fig. 7.4 Block diagram of homodyne (Zero IF/Direct-Conversion) transmitter

much less spurious products than the super-heterodyne transmitter [5]. Despite its advantages, the homodyne transmitter has a number of drawbacks, which must be dealt with for a successful implementation; these will be examined next.

7.3.1   Homodyne Transmitter Architecture Drawbacks

7.3.1.1   LO Disturbance and Its Corrections

The power amplifier output, by coupling to the local oscillator, disturbs the purity of its output frequency, Fig. 7.5. This disturbance is caused because, since the waveform at the PA output is modulated and its spectrum centered about the LO frequency, when the frequency variation reaches the LO circuit, it manifests as noise added to the LO frequency [5]. This noise, in turn, corrupts the purity of the LO signal. The corruption occurs through "injection pulling" or "injection locking" [6], which is the physical mechanism describing the shift in the frequency of an oscillator towards that of an external stimulus, Fig. 7.5c.

In particular, when the frequency of the injected noise is close to the oscillator center frequency, increases in the amplitude of the injected noise cause the LO frequency to broaden with the peak magnitude of the LO spectrum shifting towards the injected frequency until eventually the strongest signal produced by the LO circuit is that of the injected signal; this is the "locking" condition [5, 6]. It is found, in practice, that to avoid injection locking it is necessary to maintain noise levels well under 40 dB [5].

A number of approaches may be utilized to reduce LO pulling. In a first approach, the output spectrum of the PA is set far away from the LO frequency. This

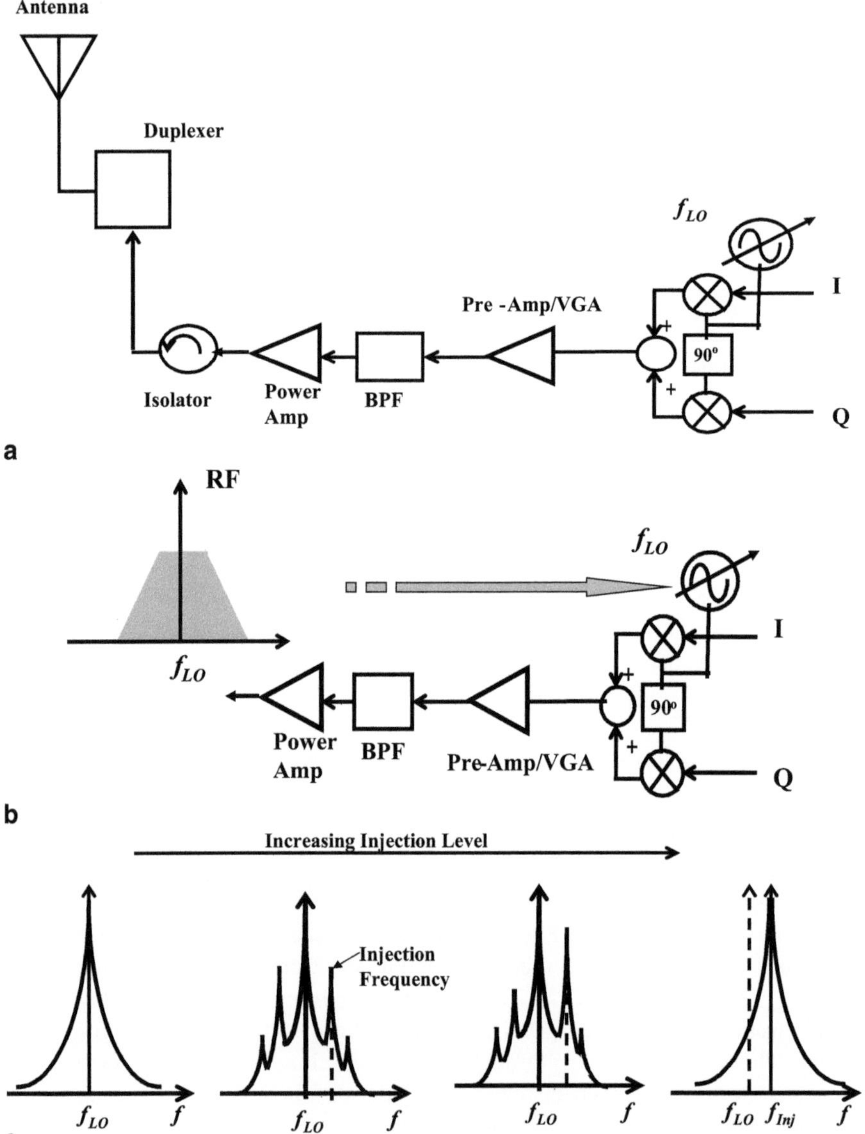

Fig. 7.5 a Direct-conversion transmitter. **b** Coupling from power amplifier output to local oscillator causing LO frequency puling. **c** Phenomena of frequency pulling by signal coupled from PA back to LO. *After* [5]

can be accomplished, in the case of quadrature up-conversion, by generating the LO frequency at an offset, i.e., at a frequency away from the intended LO frequency, and then synthesizing the actual LO frequency by adding or subtracting the offset frequency [5]. This is exemplified by Fig. 7.6, where oscillators generate frequen-

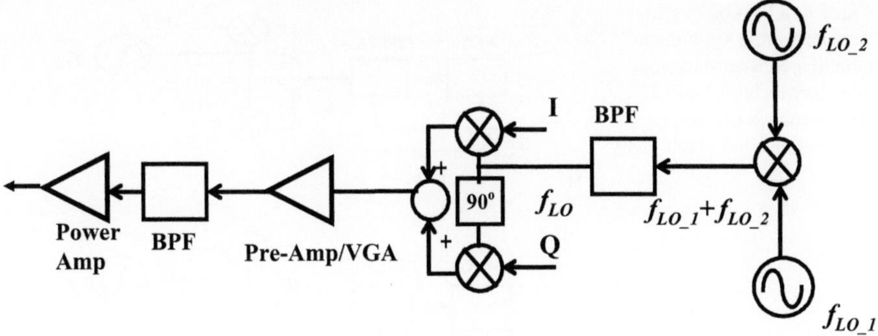

Fig. 7.6 Direct-conversion transmitter with *offset* LO to diminish frequency pulling

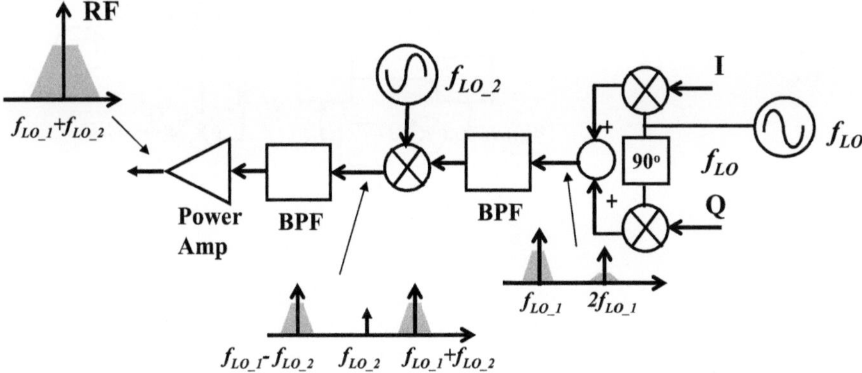

Fig. 7.7 Heterodyne transmitter utilizing two-step up-conversion. The first BPF suppresses the harmonics of the IF signal while the second removes the unwanted sideband centered around $f_{LO\_1}-f_{LO\_2}$

cies f_{LO_1} and f_{LO_2}, which are mixed and filtered to produce $f_{LO\_1}+f_{LO\_2}=f_{LO}$, such that $f_{LO}\gg f_{LO\_1}$ and $f_{LO}\gg f_{LO\_2}$. A potential difficulty with this approach, however, is that, if the filter does not suppress the mixing terms $nf_{LO\_1}\pm mf_{LO\_2}$, these will appear as spurious frequencies at the amplifier output spectrum, so the selectivity of the filter whose output is f_{LO} becomes crucial [5].

In a second approach to circumventing the problem of LO pulling in transmitters, the baseband signal is up-converted in at least two steps so that the PA output spectrum is far from the frequency of the LOs, Fig. 7.7. In this case, the baseband I and Q channels undergo quadrature modulation at a lower frequency, f_{LO_1}, and the result is up-converted to $f_{LO\_1}+f_{LO\_2}$ by mixing and band-pass filtering. As pointed out by Razavi [5], an advantage of the two-step up-conversion approach over the direct one is that, since quadrature modulation is performed at lower frequencies, a superior I and Q matching may be obtained, which results in a lower level of cross-talk between the two bit streams. Furthermore, utilization of channel filters

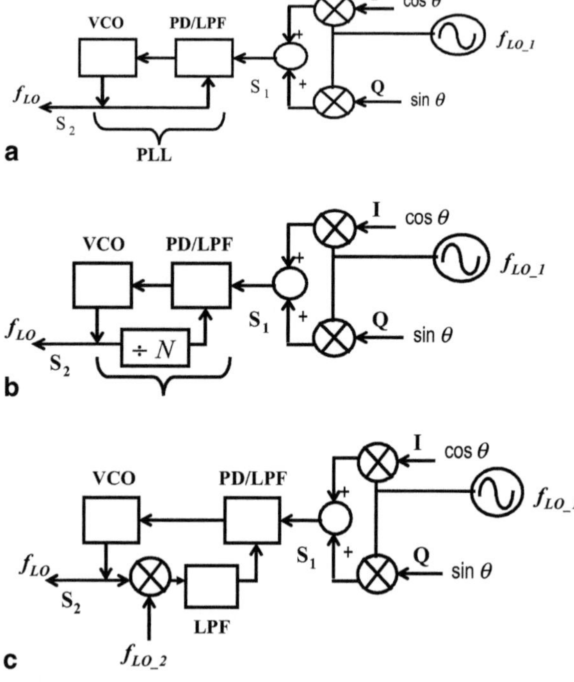

Fig. 7.8 a Up Offset-PLL Architecture. **b** Up-conversion of a constant-envelope signal by means of a PLL with feedback divider. **c** Up-conversion using offset PLL

at the first IF helps in limiting spurs and noise originating in neighboring channels [5]. On the other hand, a drawback of two-step transmitters is that, because the up-conversion process produces sidebands (both wanted and unwanted) that have the same magnitude, the filter that follows the second up-conversion must reject the unwanted sideband by at least 50 to 60 dB [5]; thus, this filter must be implemented in an expensive off-chip technology.

A third approach to reduce injection pulling is possible when a constant envelope modulation scheme is being used. In this case, the offset-PLL architecture may be employed, Fig. 7.8a [5].

This technique was invented to meet the GSM requirements for the thermal noise in the receive band [5]. In it, the PLL operates as a narrowband filter, centered at f_{LO}, suppressing the out-of-band noise generated by the I/Q modulator. If the loop bandwidth of the PLL is chosen properly, the phase information in S_1 is transferred to S_2 faithfully, while the output noise at large frequency offsets is determined predominantly by that of the VCO [5].

In some instances, e.g., at high frequencies, the realization of the phase detector is not feasible. In such cases, a divide-by-N may be inserted in the feedback path, Fig. 7.8b, which results in amplification of the phase modulation of S_1 by a factor of N when it appears in S_2 requiring higher resolution phase modulation in sin θ and cos θ. A consequence of this frequency division, however, is that the frequency steps in f_{LO_1} are also amplified by N. Therefore, whenever this scheme is used,

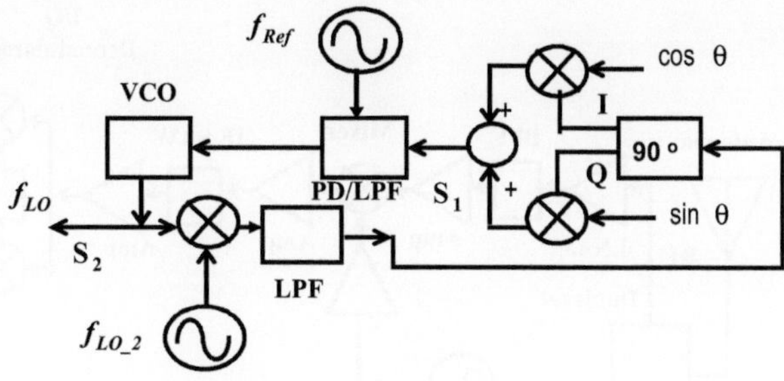

Fig. 7.9 Offset PLL including quadrature up-conversion

the necessary production of smaller steps in the synthesizer generating f_{LO_1} must provide small channel spacing, which in turn results in an undesirably long settling time [5]. These issues may be alleviated, as shown below on Fig. 7.8c by incorporating in the PLL an offset mixer driven by another oscillator so as to lower the frequency driving the PD [5].

An alternative to the offset-PLL architecture, which involves embedding the quadrature up-converter inside the loop, may be also be implemented, Fig. 7.9 [5].

In this architecture, a constant reference frequency f_{Ref} is applied to the loop, which results in minimizing the phase variation of S_1 and in modulating the phase of S_2 according to the baseband waveforms I and Q. This architecture, however, requires two high-frequency VCOs and a selective filter at the output of the offset mixer, so that its complexity is greater [5]. This complexity is compensated by the production of low out-of-band noise, which enables the operation of the transceiver without a duplexer and, thus, substantially reduces the necessary power capability of the PA, e.g., substantially less ten percent [5].

Other problems [5] germane to the homodyne transmitter arise if the system has to fulfill stringent requirements on the output power range, as is necessary in W-CDMA systems. In the first place, since most of the gain has to be realized in the baseband section, this leads to high linearity requirements for the baseband filters and the modulator. In the second place, since the LO frequency is set in the transmit band, this causes high requirements on the LO-RF isolation. And finally, I/Q phase mismatches are also an issue because, even a low error in the phase shifting network may lead to a severe degradation of the error vector magnitude (EVM) [5].

7.4 Receiver Architectures: Heterodyne

In the heterodyne receiver architecture, Fig. 7.10, the received RF signal is translated to much lower frequencies; this makes it possible to reduce the required quality factor for the channel-select filter, so its is easier to implement [4]. In the following, with reference to Fig. 7.10, we describe the architecture.

The RF front-end processes the modulated high-frequency carrier directly from the antenna. It includes part of the duplexer as the pre-selector, a low noise amplifier

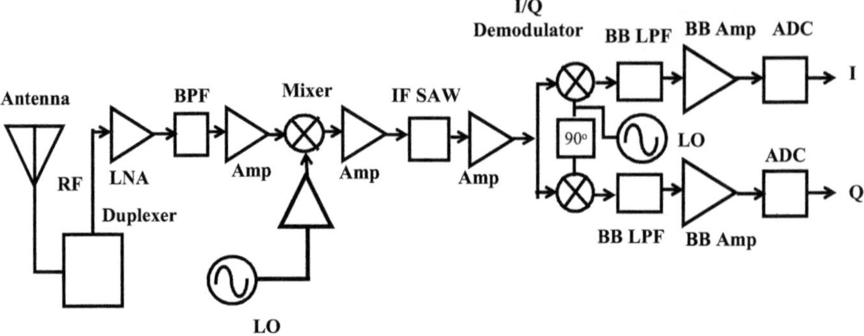

Fig. 7.10 Block diagram of super-heterodyne receiver

(LNA), an RF band pass filter (BPF), an RF amplifier as the preamplifier of the mixer, and an RF-to-IF down-converter mixer.

The LNA enables improving the reception sensitivity to a desired level, and its gain can be adjusted to compensate for deficiencies in the receiver dynamic range.

The RF BPF is usually implemented in SAW filter technology; its function is to drastically reduce leakage originating on the transmitter, as well as the image, and other interference. The SAW filter is not necessary if the rejection effected by the pre-selector to the transmission power is sufficiently large, or if the receiver is one in a half-duplex system [4].

The RF amplifier (Amp) or preamplifier preceding the mixer is endowed with a large gain to reduce the influence of the noise figure of mixer, and subsequent stages, on the overall receiver noise figure and sensitivity. This amplifier is usually necessary when the mixer is passive [4].

The mixer effects the signal frequency translation from RF to IF. It is followed by an IF amplifier and this by an IF BPF for channel selection and suppression of unwanted mixing products. This filter is usually implemented by high-selectivity SAW or crystal filters. Most of the gain of the overall receiver chain is provided by the IF amplifier which, formed by multiple stages, may produce a variable/adjustable gain.

The IF block is followed by the I/Q demodulator, which is the second frequency converter, down-converting the signal frequency from IF to BB. The demodulator contains two mixers, and it converts the IF signal into I and Q signals, namely, two BB signals phase shifted 90° with respect to each other. The 90° phase shift is implemented through a polyphase filter that shifts the phase between very-high frequency (VHF) LO signals going to the mixers in the I and Q channels.

A low-pass filter (LPF) follows the mixer in both the I and Q channels, to filter out the unwanted mixing products and to further suppress interferers.

As previously indicated, the heterodyne architecture first translates the signal band down to some intermediate frequency (IF), which is usually much lower than

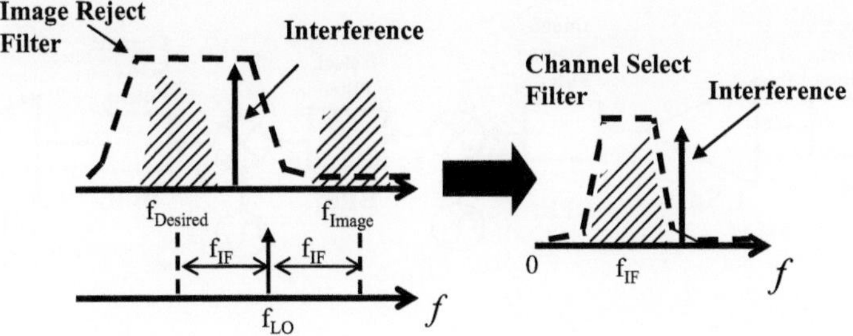

Fig. 7.11 Image rejection and channel selection for the heterodyne receiver structure

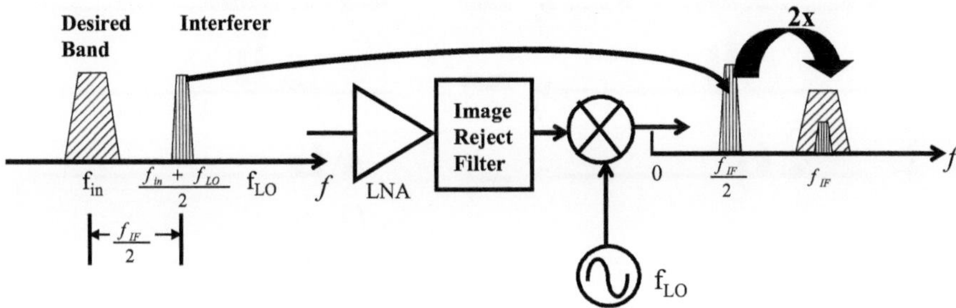

Fig. 7.12 Phenomenon of the half-IF problem

the initially received frequency band. It is at this IF that channel selection is usually effected because the low frequency reduces the demand on the performance requirements of the of the channel selection filter. In this context, the nature of the IF filter becomes the chief consideration in the design of heterodyne receivers. In particular, since the first mixer down-converts frequency bands that are located symmetrically below and above the LO center frequency, there is a need to include an image-reject filter in front of the mixer. The filter, then, is designed to have a relatively small loss in the desired band and a large attenuation in the image band, two requirements that can be simultaneously met for $2f_{IF}$ sufficiently large, Fig. 7.11 [5].

As may be appreciated from Fig. 7.11, a large IF allows using a wider image-rejection filter in front of the mixer, which is easier to realize, i.e., possesses less stringent performance requirements [5].

As seen in Fig. 7.10, the RF BPF is usually a SAW filter, whose function is to suppress the transmission leakage, the image, and other interference. Another interesting situation arises with an interferer at $(f_{wanted} + f_{LO})/2$, Fig. 7.12. If this interferer experiences second-order distortion and the LO contains a significant second harmonic, then a component at $(f_{wanted} + f_{LO}) - 2f_{LO} = f_{IF}$ arises. This phenomenon, called the "half-IF problem," is suppressed by minimizing the 2nd order distortion in the RF and IF paths, and by keeping 50 % LO duty cycle [5].

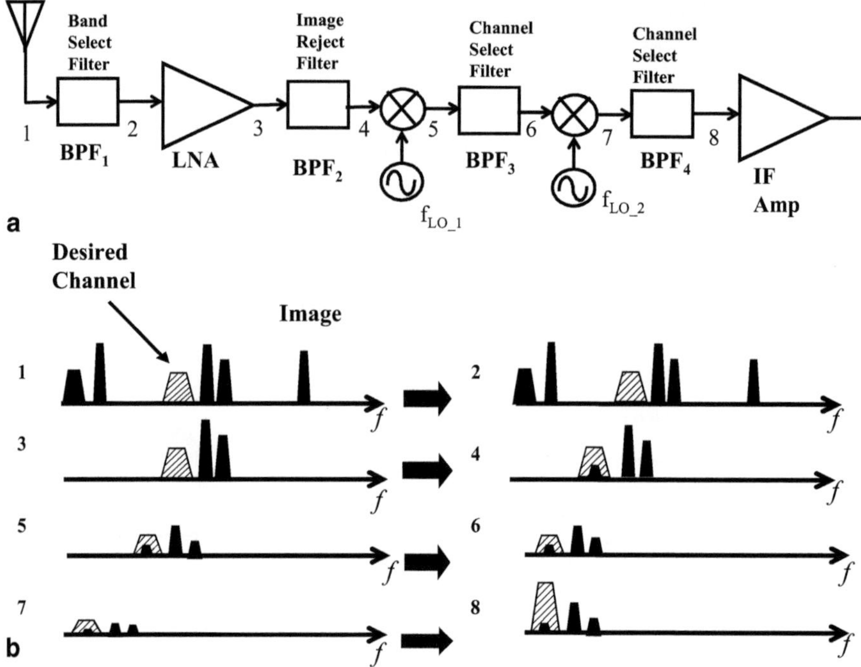

Fig. 7.13 **a** Dual IF heterodyne receiver. **b** Spectrum at numbered locations

7.5 Super-Heterodyne Receiver—Dual IF Topology

In many cases, the trade-off between sensitivity and selectivity in the single-conversion heterodyne receiver architecture is difficult to achieve successfully [5]. For instance, if the IF is high, the image can be suppressed, but complete channel selection may be impossible, and vice versa. To address this issue, the concept of heterodyning can be extended to multiple down conversions, with each mixer stage being followed by filtering and amplification. This technique effects partial channel selection at progressively lower center frequencies, thereby relaxing the Q required of each filter, Fig. 7.13 [5].

The operation of the dual-IF heterodyne receiver is now given, with respect to Fig. 7.13. First, the front-end filter selects the RF band, while also providing some image rejection ($1{\rightarrow}2$). Secondly, upon effecting amplification and image reject filtering, the spectrum at point **3** is obtained. Thirdly, a mixer translates the desired channel and the adjacent interferers to the first IF (**4**). Fourthly, the partial channel selection effected by BPF_3 permits the use of a second mixer with reasonable linearity (**5, 6**). Fifthly, the spectrum is translated to the second IF, and BPF_4 suppresses the interferers to acceptably low levels (**7, 8**) [5].

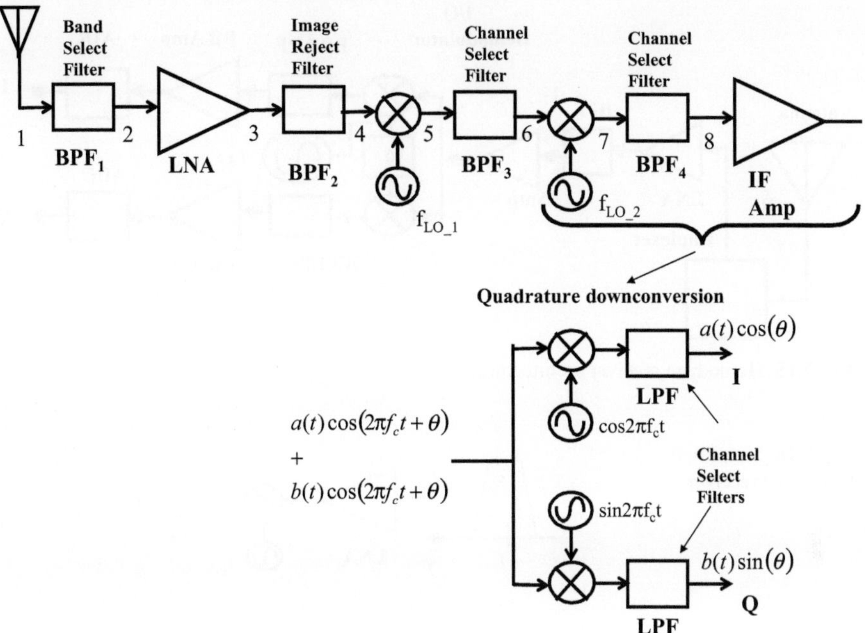

Fig. 7.14 Quadrature down-conversion

Subsequent processing, after the second IF, depends on the nature of the system. In analog FM systems, this latter IF frequency is demodulated to extract the baseband signal [5]. On the other hand, in digital modulation systems, the second conversion usually generates both in-phase (I) and quadrature (Q) components of the signal, while translating the spectrum to zero frequency, Fig. 7.14 [5].

7.6 Homodyne (Zero IF/Direct-Conversion) Receiver

Examining the heterodyne receiver front end, we see that the function of the BPF in the duplexer is to suppress the leakage power of the transmission and other out of receiver band interference, e.g., the image. There is no image problem, however, for the direct-conversion receiver, since it does not have an image band. The received signal, after pre-selection by the duplexer, is amplified by an LNA, and is further filtered by an RF filter, Fig. 7.15.

Due to the leakage from the transmitter, and to control the LO-RF leakage-related self-mixing, the rejection of this filter needs to be higher than that required in the heterodyne receiver; this higher rejection also allows the relaxation of the second-order distortion requirement of the down-converter [5].

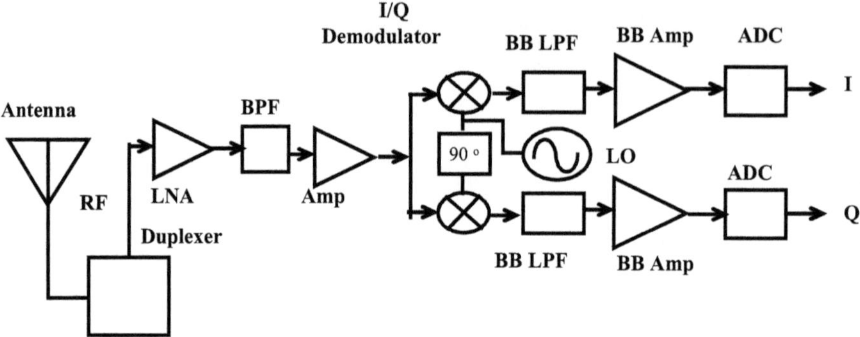

Fig. 7.15 Homodyne receiver architecture

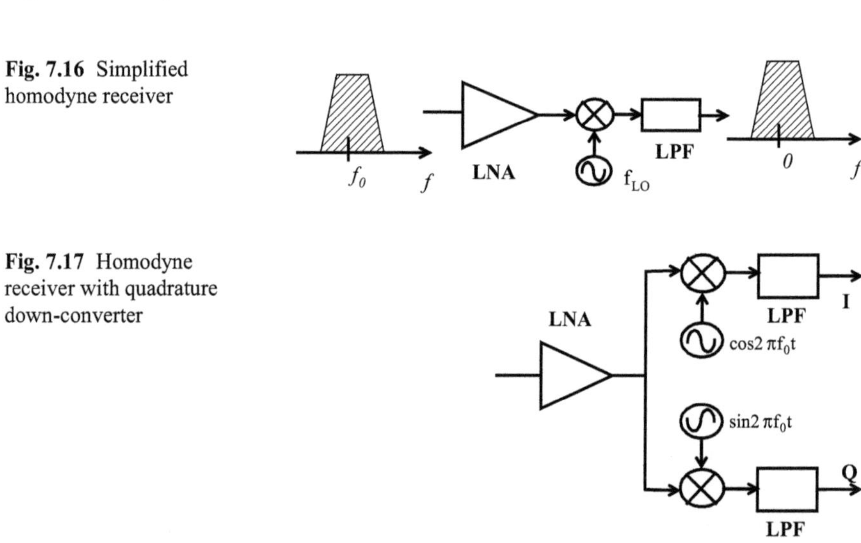

Fig. 7.16 Simplified homodyne receiver

Fig. 7.17 Homodyne receiver with quadrature down-converter

We continue discussing the homodyne receiver by focusing on a simplified block diagram, Fig. 7.16.

In the homodyne or, zero-IF receiver, the LO frequency is equal to the received input carrier frequency. Therefore, channel selection requires only a low-pass filter with relatively sharp cutoff characteristics [5]. Since it overlaps positive and negative parts of the input spectrum, the circuit in Fig. 7.16 operates properly only with double-sideband AM signals. For frequency and phase modulated signals, the down conversion must provide quadrature outputs to avoid loss of information, this is effected by the circuit on Fig. 7.17. This is because the two sides of FM or QPSK spectra carry different information and must be separated into quadrature phases in translation to zero frequency [4, 5].

The homodyne receiver has a number of advantages over the heterodyne receiver. In particular, the image problem is circumvented because $f_{IF} = 0$, therefore, no image filter is required, the LNA need not drive a 50 Ω load, and the IF SAW filter and subsequent down conversion stages are replaced with low-pass filters and baseband filters that are amenable to monolithic integration [5].

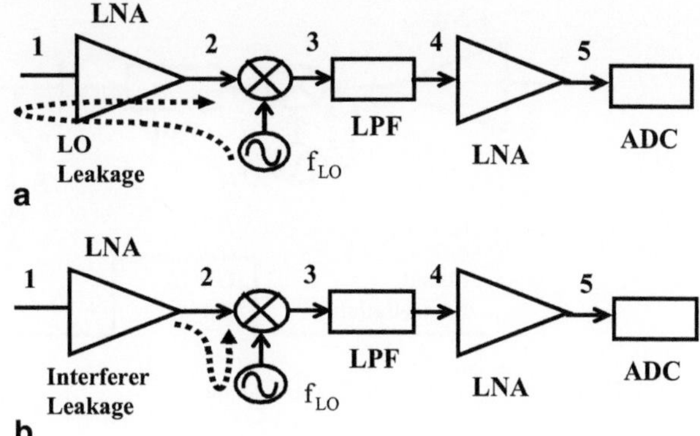

Fig. 7.18 Source of DC-offsets in direct-conversion receiver. **a** LO-RF leakage (This phenomenon is called "self-mixing"). **b** Interferer leakage

7.6.1   Homodyne Receiver—Drawbacks

7.6.1.1   DC Offsets

Since in a homodyne topology the down-converted band extends to zero frequency, extraneous offset voltages can corrupt the signal and saturate the subsequent stages. These offsets originate due to the finite (low) isolation between the LO and the LNA input, which results in a signal at the LNA input which is then mixed again with the LO signal, consequently producing a DC component at point 4, Fig. 7.18a.

A similar situation occurs if a large interferer leaks from the LNA or mixer into the LO port and is multiplied by itself, Fig. 7.18b. An exercise estimating the DC Offset has been provided by Razavi [5] as follows. Assuming the LO power to be 0 dBm or 0.63 Vpp in a 50 Ω system, and that it experiences a 60 dB attenuation as it couples to point 1. Then, if the gain of the LNA/mixer combination is 30 dB, the offset produced at the output of the mixer is about 10 mV. Since the level of the desired signal can be as low as 30 mV*rms*, if directly amplified by 50 to 70 dB, the offset voltage saturates the following circuits, thereby precluding the amplification of the desired signal. The problem worsens if self-mixing varies with time, which case can occur if the LO leaks to the antenna, is radiated and subsequently reflected back from moving objects to the receiver. Under these circumstances, it may be difficult to distinguish the time-varying offset from the actual signal, and offset cancellation techniques are necessary.

7.6.1.1.1   Offset Cancellation Approaches

The following techniques are normally applied to cancel the DC offset [5]. The first one is based on *DC-free coding*. In this technique, the baseband signal in the transmitter is encoded so that, after modulation and down-conversion, it contains very little energy near DC. In the second technique, called *Measure and Cancel DC Offset*, the idle time intervals in digital wireless standards is exploited to carry out

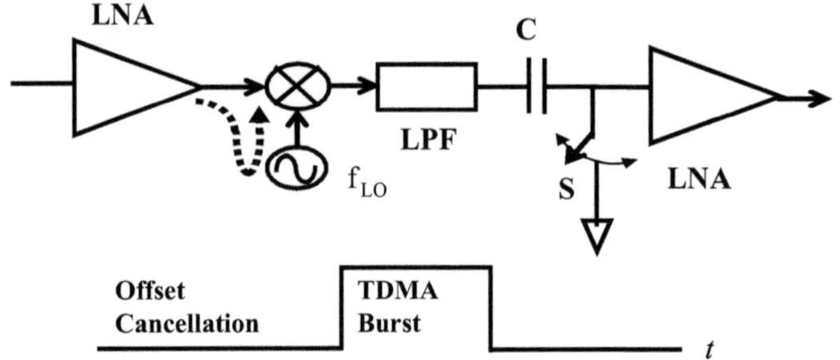

Fig. 7.19 Offset cancellation in TDMA system

offset cancellation, Fig. 7.19. In the figure, a capacitor stores the offset between consecutive TDMA bursts, while introducing a virtually zero corner frequency during the reception of data. For a typical TDMA frame of a few milliseconds, performing the offset cancellation exercise with sufficient frequency can compensate variations due to moving objects [5]. In Fig. 7.19, a capacitor stores the offset between consecutive TDMA bursts, while introducing a virtually zero corner frequency during the reception of data.

7.6.1.2   I/Q Mismatch

Another drawback concomitant with the homodyne receiver is *I/Q mismatch*. This is rooted on that fact that for phase and frequency modulation schemes, a homodyne receiver must incorporate quadrature mixing, which requires shifting the LO output by 90°. Unfortunately, the errors in the nominally 90° phase shift and mismatches between the amplitudes of the I and Q signals corrupt the down-converted signal constellation, thereby raising the bit error rate [5, 7]; Fig. 7.20.

This error is understood as follows [5]. If the received signal is:

$$x_{in}(t) = a\cos\omega_c t + b\sin\omega_c t \tag{7.1}$$

where a and b are -1 or $+1$. If the I and Q phases of the LO signals are:

$$x_{LO,I}(t) = 2\left(1+\frac{\varepsilon}{2}\right)\cos\left(\omega_c t + \frac{\theta}{2}\right) \tag{7.2}$$

and

$$x_{LO,Q}(t) = 2\left(1-\frac{\varepsilon}{2}\right)\sin\left(\omega_c t - \frac{\theta}{2}\right) \tag{7.3}$$

where ε and θ represent amplitude and phase errors, respectively. Multiplying $x_{in}(t)$ by the two LO phases and low-pass filtering the results, one obtains the following baseband signals,

Fig. 7.20 I/Q mismatch contributions by various stages

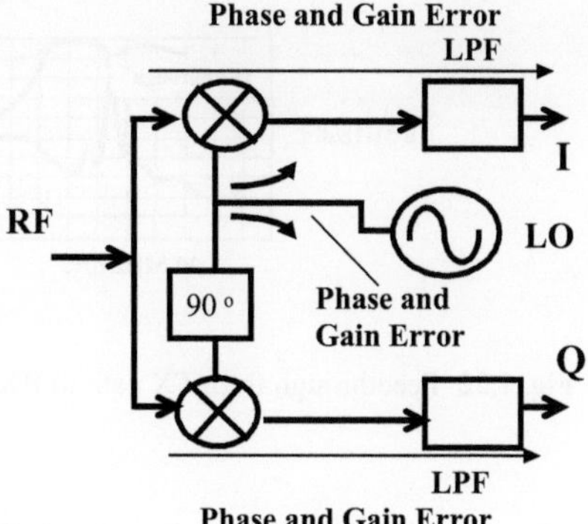

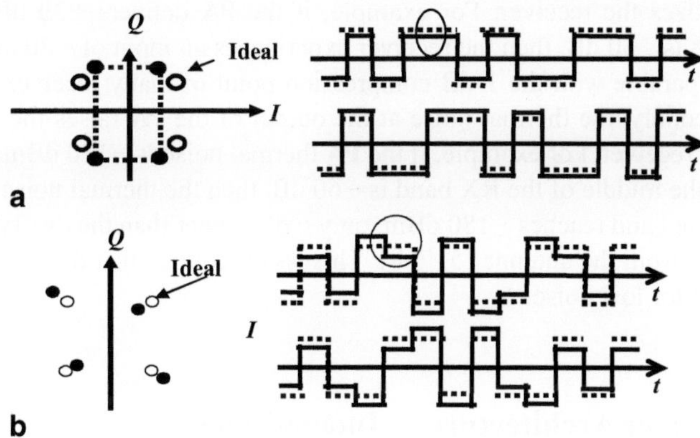

Fig. 7.21 a Effect of gain error on QPSK constellation. **b** Effect of phase error on QPSK constellation

$$x_{BB,I}(t) = a\left(1+\frac{\varepsilon}{2}\right)\cos\left(\frac{\theta}{2}\right) - b\left(1+\frac{\varepsilon}{2}\right)\sin\left(\frac{\theta}{2}\right) \qquad (7.4)$$

and

$$x_{BB,Q}(t) = -a\left(1-\frac{\varepsilon}{2}\right)\sin\left(\frac{\theta}{2}\right) + b\left(1-\frac{\varepsilon}{2}\right)\cos\left(\frac{\theta}{2}\right) \qquad (7.5)$$

The effects of gain and phase errors on a QPSK constellation is shown in Fig. 7.21.

7.7 Transmitter Leakage

The function of the receiver BPF in the duplexer is to suppress the leakage power of the transmission and other out of receiver band interference [7]. This effect arises simply because of the finite attenuation of the TX filter in the receive band – only

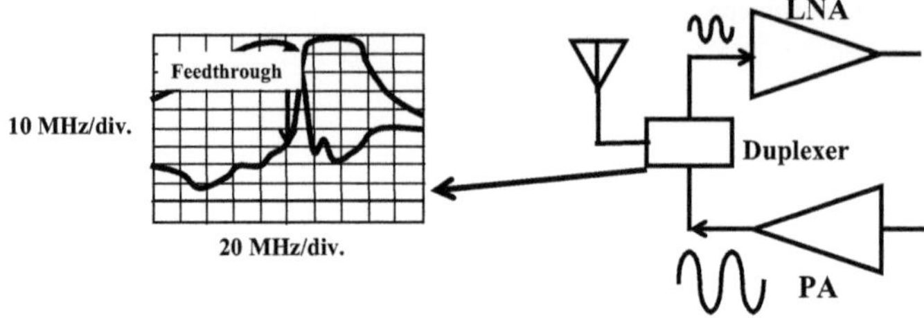

Fig. 7.22 Feedthrough from TX path to RX path

45 to 50 dB at the edge, Fig. 7.22. The feedthrough of the **PA** output to the **LNA** input creates two difficulties. Firstly, the large signal generated by the PA heavily desensitizes the receiver. For example, if the **PA** delivers $+\mathbf{30}$ dBm and the feedthrough is -50 dB, then the receiver experiences an input of -20 dBm, a level that is comparable with the 1-dB compression point of many receiver designs [5, 8]. And secondly, the thermal noise at the output of the **PA** raises the input noise floor of the receiver. For example, if the **PA** thermal noise **is** -120 dBm/Hz and the leakage in the middle of the RX band is -60 dB, then the thermal noise introduced in the receive band reaches -180 dBm, only 6 dB higher than the available thermal noise power from the antenna [5, 7, 8]. This issue requires that the entire **TX** path be designed for low noise.

7.8   Receiver Architectures—Image-Reject

The trade-offs that govern the use of image-reject filters in heterodyne architectures have motivated seeking other techniques for suppressing the image. These include the Hartley and the Weaver architectures, which are discussed next.

7.8.1   Hartley Architecture

The Hartley receiver architecture is shown in Fig. 7.23 [4, 5, 8]; its operation is as follows. The circuit mixes the RF input with the quadrature phases of the LO, $\sin\omega_{LO}t$ and $\cos\omega_{LO}t$, low-pass filters the resulting signals, and shifts one by $90°$ before adding them together. Mathematically, this is expressed as follows. Assuming the input signal is,

$$x(t) = \underbrace{A_{RF} \cos \omega_{RF} t}_{\text{Desired}} + \underbrace{A_{im} \sin \omega_{im} t}_{\text{Image}} \qquad (7.6)$$

Fig. 7.23 Hartley receiver
architecture

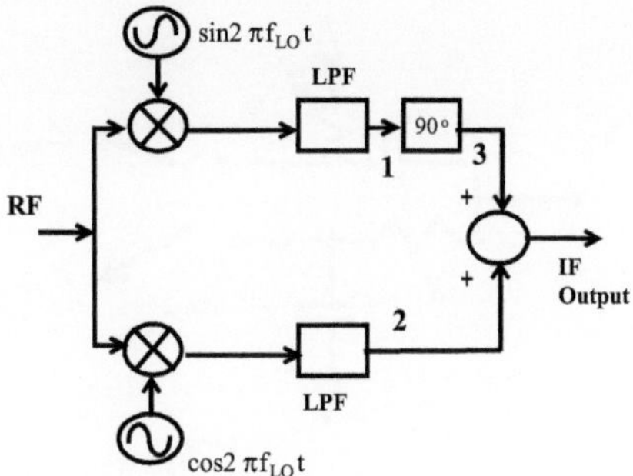

Then, assuming: $\omega_{RF}-\omega_{LO}=\omega_{LO}-\omega_{im}$ it follows that multiplying $x(t)$ by the LO phases and neglecting high-frequency components, one obtains the following at points 1 and 2.

$$x_1(t) = \frac{A_{RF}}{2}\sin(\omega_{LO}-\omega_{RF})t + \frac{A_{im}}{2}\sin(\omega_{LO}-\omega_{im})t \qquad (7.7)$$

and

$$x_2(t) = \frac{A_{RF}}{2}\cos(\omega_{LO}-\omega_{RF})t + \frac{A_{im}}{2}\cos(\omega_{LO}-\omega_{im})t \qquad (7.8)$$

Since $\sin(x)=-\sin(-x)$, one can apply this to the first term of $x_1(t)$ and write:

$$x_1(t) = -\frac{A_{RF}}{2}\sin(\omega_{RF}-\omega_{LO})t + \frac{A_{im}}{2}\sin(\omega_{LO}-\omega_{im})t \qquad (7.9)$$

Also, using the identity $\sin(x-90°)=-\cos(x)$, one can impose the effect on $x_1(t)$ of the $90°$ phase shift, to obtain:

$$x_3(t) = \frac{A_{RF}}{2}\cos(\omega_{RF}-\omega_{LO})t - \frac{A_{im}}{2}\cos(\omega_{LO}-\omega_{im})t \qquad (7.10)$$

Then, adding $x_3(t)$ and $x_2(t)$, one obtains:

$$x_{IF}(t) = A_{RF}\cos(\omega_{RF}-\omega_{LO})t \qquad (7.11)$$

In the Hartley receiver, therefore, the signal components have the same polarity, but the image components have opposite polarity [4, 5, 8].

The practical implementation of the Hartley receiver relies on that of the 90° phase shift. This is realized with a combination of two RC circuits, one producing $+45°$ and the other $-45°$, a split-phase approach, Fig. 7.24a. The final image-reject Hartley receiver is as shown in Fig. 7.24b.

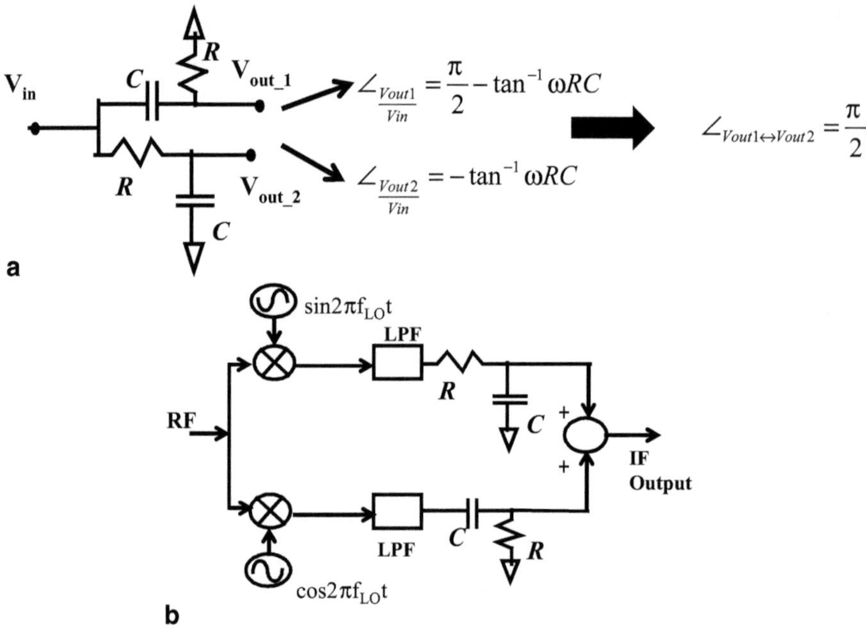

a

b

Fig. 7.24 **a** Practical implementation of 90° phase shift. **b** Image-reject receiver with split phase shift stages

7.8.1.1 Hartley Architecture—Drawbacks

The fundamental drawback of the Hartley receiver is its sensitivity to *mismatches* [4, 5, 8] In particular, if the LO phases are not in exact quadrature, or the gains and phase shifts of the upper and lower paths are not identical, then the cancellation is incomplete and the image corrupts the down-converted signal. In other words, suppose the LO phases in Fig. 7.23 are given by,

$$x_1(t) = \frac{A_{LO}A_{RF}}{2}\sin(\omega_{LO} - \omega_{RF})t + \frac{A_{LO}A_{im}}{2}\sin(\omega_{LO} - \omega_{im})t \qquad (7.12)$$

and

$$x_2(t) = \frac{(A_{LO} + \varepsilon)A_{RF}}{2}\cos(\omega_{LO} - \omega_{RF})t + \theta + \frac{(A_{LO} + \varepsilon)A_{im}}{2}\cos(\omega_{LO} - \omega_{im})t + \theta \qquad (7.13)$$

then,

$$x_3(t) = A_{LO}\left[\frac{A_{RF}}{2}\cos\left(\omega_{RF} - \omega_{LO}\right)t - \frac{A_{im}}{2}\cos\left(\omega_{LO} - \omega_{im}\right)t\right] \qquad (7.14)$$

Therefore, adding $x_2(t)$ and $x_3(t)$ yields both the desired signal and a fraction of the down-converted image,

$$x_{sig}(t) = \frac{(A_{LO} + \varepsilon)A_{RF}}{2}\cos\left[(\omega_{LO} - \omega_{RF})t + \theta\right] + \frac{(A_{LO} + \varepsilon)A_{im}}{2}\cos\left[(\omega_{LO} - \omega_{im})t + \theta\right]$$

(7.15)

$$x_{im}(t) = \frac{(A_{LO} + \varepsilon)A_{RF}}{2}\cos\left[(\omega_{LO} - \omega_{im})t + \theta\right] - \frac{A_{LO}A_{im}}{2}\cos(\omega_{LO} - \omega_{im})t \quad (7.16)$$

The mismatch is characterized by the image-to-signal ratio at the output, given by,

$$\left.\frac{P_{im}}{P_{sig}}\right|_{out} = \frac{A_{im}^2}{A_{RF}^2} \cdot \frac{(A_{LO} + \varepsilon)^2 - 2A_{LO}(A_{LO} + \varepsilon)\cos\theta + A_{LO}^2}{(A_{LO} + \varepsilon)^2 + 2A_{LO}(A_{LO} + \varepsilon)\cos\theta + A_{LO}^2}$$

(7.17)

Defining the term $\dfrac{P_{im}}{P_{sig}} / \dfrac{A_{im}^2}{A_{RF}^2}$ as *image rejection ratio* (*IRR*), we can write,

$$IRR = \frac{A^2 - 2AB\cos\theta + B^2}{A^2 + 2AB\cos\theta + B^2}$$

(7.18)

where $A = A_{LO}$ and $B = A_{LO} + \varepsilon$ embody the two different gains of the channels. For $\varepsilon \ll A_{LO}$ and $\theta \ll 1$ rad, this reduces to:

$$IRR = \frac{(\Delta A / A)^2 + \theta^2}{4}$$

(7.19)

With this expression, mismatches in mixers, LPFs, and the two ports of the adder can be lumped into $\Delta A/A$ *and* θ, where $\Delta A/A = \varepsilon/A_{LO}$ denotes the mismatch. For typical matching in integrated circuits, the image suppression falls in the 30–40 dB range, reflecting an overall gain mismatch of 0.2–0.6 dB and 1–5 ° phase imbalance [4, 5, 8].

Another source of mismatch results from error in the circuit effecting the 90° phase shift operation. In particular, it can be shown that if $R \to R + \Delta R$ and $C \to C + \Delta C$ in the RC circuits,

$$\frac{\Delta A}{A} = \frac{(R + \Delta R)(C + \Delta C)\omega - 1}{\sqrt{1 + (R + \Delta R)^2(C + \Delta C)^2\omega^2}} \div \frac{1}{\sqrt{1 + R^2C^2\omega^2}}$$

(7.20)

which, at $\omega RC \sim 1$, gives,

$$\frac{\Delta A}{A} \approx \frac{\Delta R / R + \Delta C / C}{\sqrt{2 + (\Delta R / R)^2(\Delta C / C)^2}} \div \frac{1}{\sqrt{2}}$$

$$\approx \frac{\Delta R}{R} + \frac{\Delta C}{C}$$

(7.21)

For $\Delta R/R = 20\%$ →IRR$=20$ dB. In addition to the mismatch due to errors in component values, gain mismatch of the two phase shift stages occurs also due to frequency deviation, since the exact image cancellation only occurs at $\omega_{IF} = 1/(RC)$. If the channel BW is not much less than ω_{IF}, then the *IRR* degrades substantially near the edges of the channel [5]. It turns out that the level of image suppression that is adequate is of the order of 60–70 dB, which is rather difficult to achieve in practice [4, 5].

7.8.2 *Weaver Architecture*

To overcome the difficulty in attaining adequate image suppression, the Weaver receiver architecture replaces the $90°$ stage by a second quadrature mixing operation to perform the same function, Fig. 7.25. The detailed operation of this receiver is as follows [5].

With respect to Fig. 7.24, the spectrum at **1** is convolved with $j[\delta(f + f_{LO2}) - \delta(f - f_{LO2})]/2$ to yield at point **3** the translated replicas with no factor j. The spectrum at **2** is convolved with $[\delta(f + f_{LO2}) + \delta(f - f_{LO2})]/2$ to yield the spectrum at point **4**. Then, subtracting the spectrum at point **3** from that at point **4**, we note that the replicas of the image that fall in the band of interest cancel each other, yielding the desired signal with corruption. Since the down converted spectrum still contains the image at $+f_{LO2} + f_{IF}$ and $-f_{LO2} - f_{IF}$, an output LPF is required to select only the desired band.

In the weaver receiver, the two LO frequencies f_{LO1} and f_{LO2} can be chosen in different ways. For the case illustrated in Fig. 7.26, where the final spectrum is not centered at zero frequency, the second mixing operation entails the problem of "secondary image."

Assume, for instance, that the input spectrum contains an interferer at $2f_{LO2} - f_{in} + 2f_{LO1}$, Fig. 7.25 (top). Then, upon the first down conversion, the interferer appears at $2f_{LO2} - f_{in} + f_{LO1}$, Fig. 7.25 (center), that is, as the image of the signal with respect to f_{LO2}. In the second down conversion, the interferer is not canceled because it is originally on the same side of f_{LO1} as the desired signal, Fig. 7.25 (bottom). For this reason, the low-pass filters after the first down conversion must be replaced by band-pass filters to suppress the secondary image [5].

In conclusion, the Hartley and Weaver architectures share one problem, namely, incomplete image rejection due to gain and phase mismatch. The Weaver circuit is free from the gain imbalance, but suffers from the secondary image problem if the second down-conversion translates the spectrum to a nonzero frequency [5, 8].

7.8.3 *Digital-IF Receiver*

In this architecture, low-frequency operations, such as the second set of mixing in the dual-IF heterodyne architecture, can be performed more efficiently in the digital

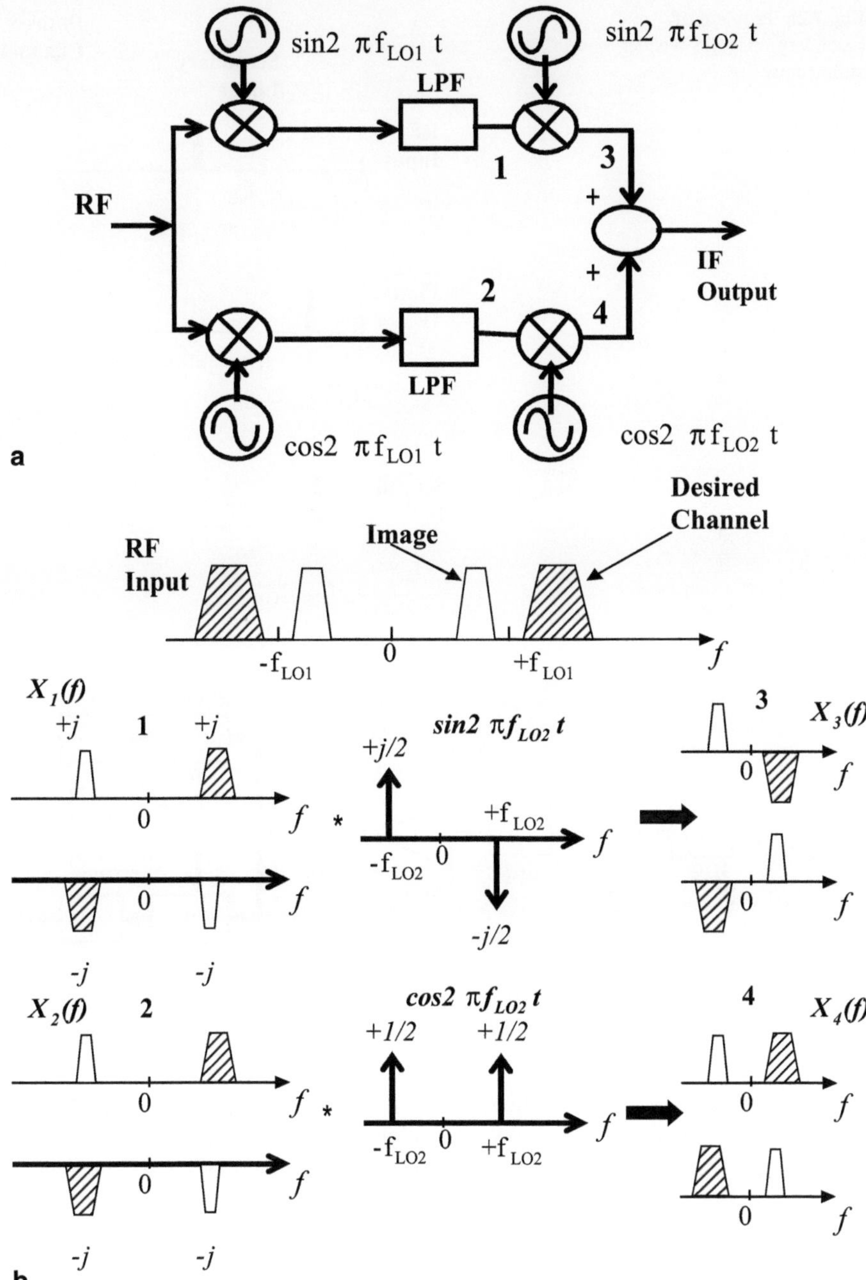

Fig. 7.25 **a** Weaver image-reject receiver. **b** Graphical Analysis

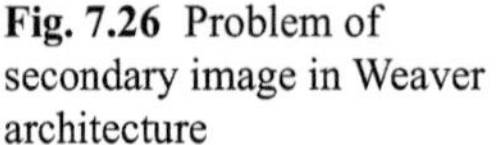

Fig. 7.26 Problem of
secondary image in Weaver
architecture

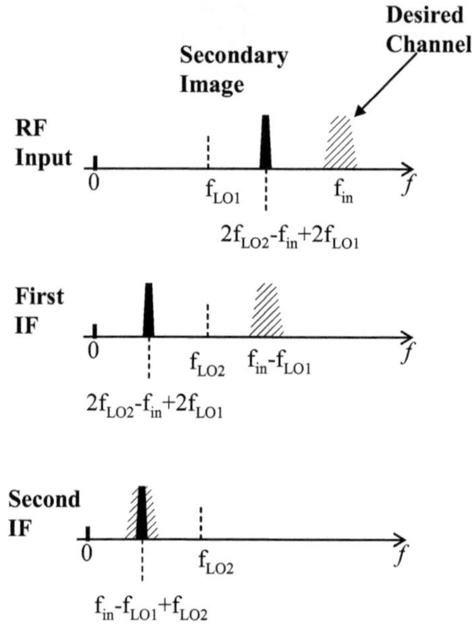

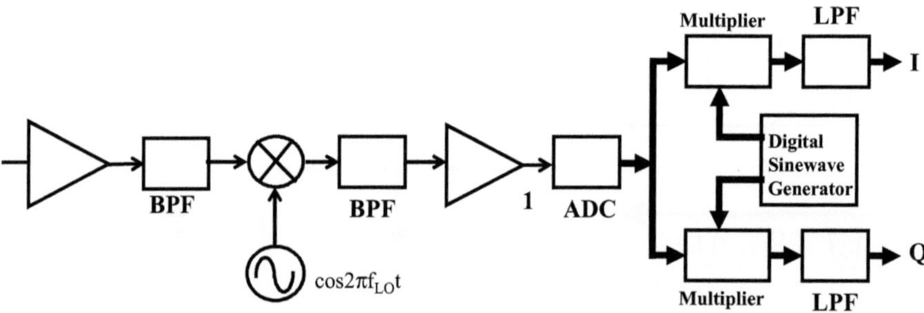

Fig. 7.27 Digital-IF receiver

domain. In Fig. 7.27, the first IF signal is digitized, mixed with the quadrature
phases of a digital sinusoid and low-pass filtered to yield the quadrature baseband
signals. Note that digital signal processing avoids the problem of I and Q mismatch.

The principal issue in the digital-IF approach is the performance required of the
A/D converter. In particular, since the signal level at point **1**, Fig. 7.26, is about a
few hundred microvolts, the quantization and thermal noise of the ADC must not
exceed a few tens of microvolts [5]. Therefore, if the first IF BPF cannot adequately
suppress adjacent interferers, the nonlinearity of the ADC must be sufficiently small
to minimize corruption of the signal by intermodulation, the ADC dynamic range
must be wide enough to accommodate variations in the signal level due to path loss
and multipath fading, and the ADC must achieve an input BW commensurate with

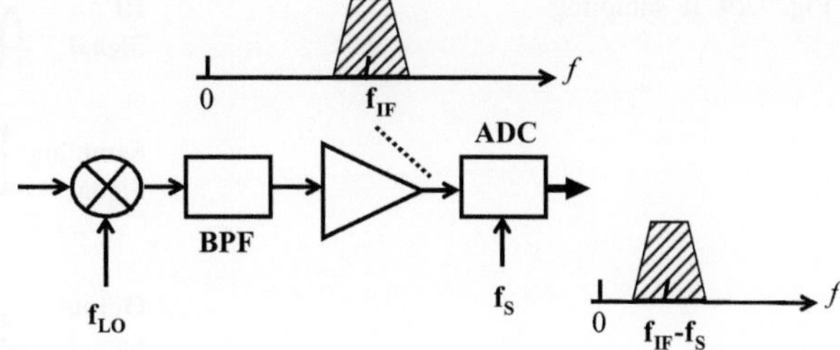

Fig. 7.28 Sampling IF architecture

the value of IF while consuming a reasonable amount of power [5]. These requirements make it difficult to employ a Nyquist-rate ADC in the digital-IF architecture. However, the demanding Nyquist-rate ADC performance may be alleviated by noting that most ADCs incorporate sample-and-hold circuits and hence can perform *down conversion* [5]. This motivates the sampling IF Architecture, discussed next.

7.8.4   Sampling IF Receiver

In the sampling IF receiver architecture, Fig. 7.28, the ADC samples the signal at a rate slightly below f_{IF}.

The spectrum of the down-converted, digitized signal thus lies around $f_{IF} - f_S$. This operation is followed by quadrature mixing and filtering to translate the spectrum to the baseband, however, even by relaxing the ADC sampling rate by a factor of 2 this technique still requires a combination of prohibitively high speed and high linearity. Nevertheless, it is used in base stations where many channels must be received and processed simultaneously [5].

In this technique, the first LO is NOT assumed to be near the RF band. In this approach, the RF is sampled at a much lower rate over many cycles. This is acceptable because narrowband signals exhibit only a small change from one carrier cycle to the next. The idea is that a bandpass signal with BW Δf can be translated to a lower band if sampled at a rate equal to $2\Delta f$, Fig. 7.29.

The idea is that a bandpass signal with BW Δf can be translated to a lower band if sampled at a rate equal to $2\Delta f$. For ideal sampling, the operation creates replicas of the spectrum with no aliasing, Fig. 7.30. Due to the large reduction in sampling rate, the use of sub-sampling can simplify the design of the local oscillator.

The drawback of the IF sub-sampling approach is aliasing of noise. In particular, the sampler equivalent circuit may be represented as an RC circuit with BW at least equal to the input frequency f_0 so it experiences minimum attenuation.

The resistor of this circuit contributes noise components from DC to beyond f_0, i.e., the noise is NOT narrow band. Therefore, sampling folds the RF noise spectrum over low frequencies and it multiplies the noise power. Sub-sampling by a factor of m multiplies the down converted noise power of the sampling circuit by a factor of $2m$ [5].

Fig. 7.29 IF sampling

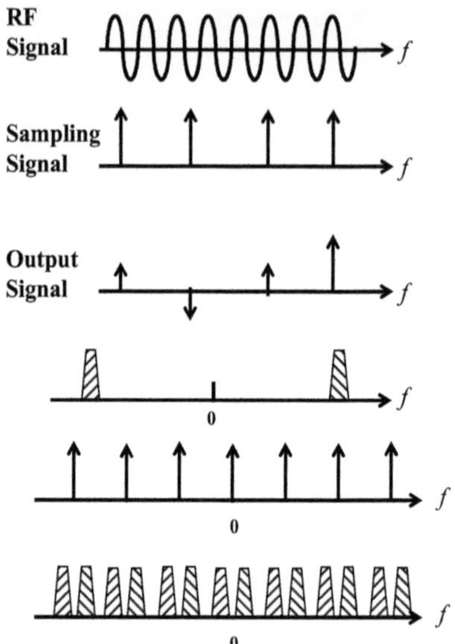

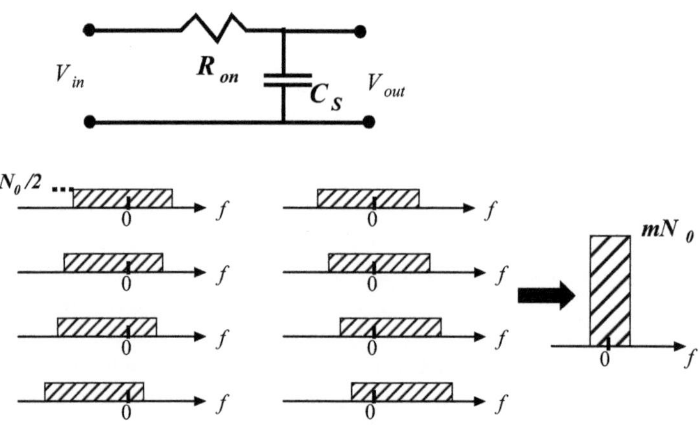

Fig. 7.30 (*Top*) Equivalent circuit of sampler. (*Bottom*) Aliasing of noise in sub-sampling

7.9   Oscillators—Phase Noise, Oscillator Pulling and Pushing

As should be apparent by now, oscillators are a key component in communications systems, providing, among other functions, the high-frequency carrier transmitted and the reference signal used in the demodulation process at the receiver. In this section we address some practical aspects of oscillators, namely, their performance and certain aspects that impact it.

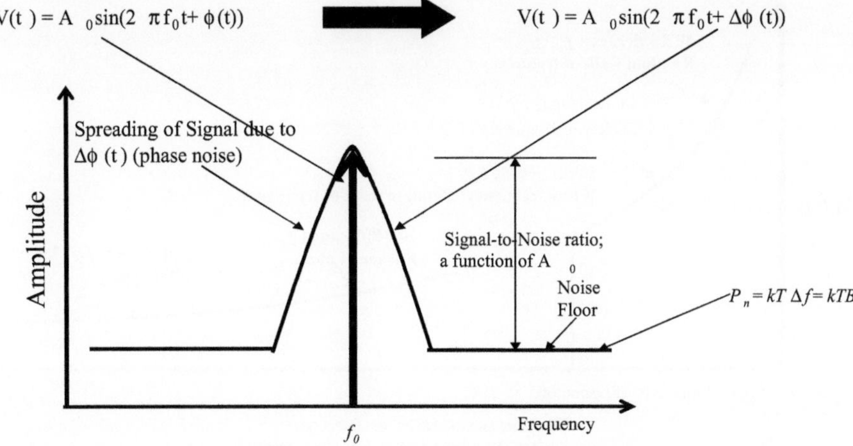

Fig. 7.31 Frequency domain signal (*spectrum*) of V(t). Where: A_0 nominal peak voltage, f_0 represents the nominal fundamental frequency, t is the time, $\varDelta\phi$ (t) is the random deviation of phase from nominal—"phase noise," and $\varDelta\phi$ and B both represent the bandwidth in which the measurement is made, in Hertz. *After* [9]

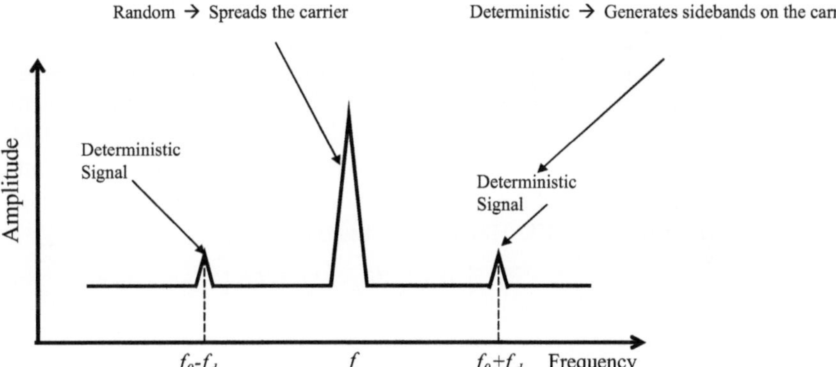

Fig. 7.32 Frequency of $V(t) = A_0 \ sin[2\pi f_0 t + \varDelta\phi(t) + m_d sin(2\pi f_d t)]$ showing deterministic signal, where m_d represents the amplitude of the deterministic signal, which is phase modulating the carrier, and f_d represents the frequency of the deterministic signal. *After* [9]

7.9.1   Phase Noise

In an oscillator, phase noise is the rapid random fluctuations in the phase component of the output signal, Fig. 7.31. The noise characteristics of the output signal of an oscillator are depicted in Fig. 7.32 [7]. As can be seen, noise on a carrier can be separated into two categories, namely, random fluctuations, which manifests itself as a spreading of the carrier, and deterministic, which generates/adds sidebands to the carrier.

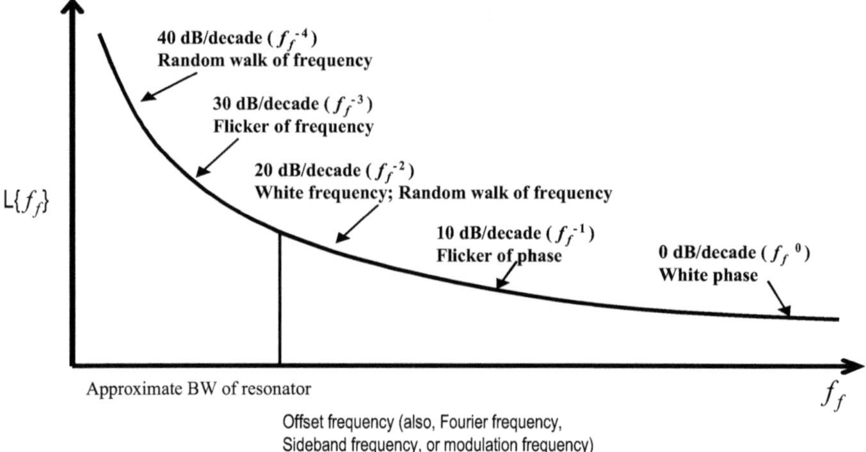

Fig. 7.33 Typical SSB phase noise plot of oscillator vs. Offset from carrier. *After* [9]

Noise has an infinite bandwidth, therefore, the greater the BW of the instrument being used to measure a carrier frequency with noise, the greater the noise it measures. For this reason, the industry has settled on a *correlation BW* for phase noise measurements of 1 Hz, known as the normalized frequency.

The noise spectrum of a signal is symmetrical around the carrier frequency and, therefore, it is necessary to specify only one side. The one-sided spectrum is called *a single sideband (SSB) spectrum*; the spectral purity of a signal being completely quantified by its SSB phase noise. The SSB plot is assigned the script L(f) and is defined as one-half the sum of both sidebands, Fig. 7.33. L(f) has units of decibels below the carrier per Hertz (dBc/Hz) and is defined as,

$$L\{f\} = 10\log\left[\frac{P_{sideband}\left(f_0 + \Delta f, 1Hz\right)}{P_{carrier}}\right] \tag{7.22}$$

where $P_{sideband}$ $(f_0 + \Delta f,\ 1\ Hz)$ represents the signal power at a frequency offset of Δf away from the carrier with a measurement BW of 1 Hz.

7.9.2 *Jitter and Random Jitter*

The characterization of the noise performance of an oscillator, in the time domain, is known as *jitter*. Phase noise and jitter are two linked quantities associated with a noisy oscillator, in particular, as the phase noise increases in the oscillator, so does the jitter. The jitter is a variation in the zero-crossing times of a signal, or a variation in the period of the signal. It has two major components: random jitter (which has a Gaussian probability density function-PDF) and deterministic jitter (non-Gaussian PDF).

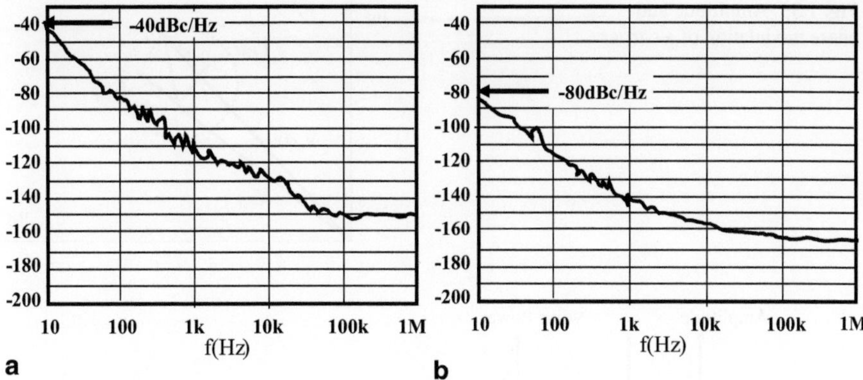

Fig. 7.34 a SSB phase noise plot of a commodity (*cheap*) clock. **b** SSB phase noise plot of a true ultralow (*expensive*) phase noise oscillator. *After* [9]

The random jitter originates from random phase noise; it results from accumulation of random processes, including: thermal noise, flicker noise, shot noise, etc.

The deterministic jitter, on the other hand, originates from deterministic noise. This is exemplified by [9]: (1) Periodic jitter or sinusoidal noise, which is caused by power supply feedthrough; (2) Intersymbol interference, which derives from channel dispersion of filtering; (3) Duty cycle distorsion, which derives from asymmetric rise/fall times; (4) Sub-harmonic(s) of the oscillator, which derives from straight multiplication oscillator designs; (5) Uncorrelated periodic jitter, which derives from crosstalk by other signals; and (6) Correlated periodic jitter. Typical phase noise performance for inexpensive and expensive oscillators are shown in Fig. 7.34 [9].

7.9.3   Analysis of Phase Noise

The phase noise of an oscillator can be the result of amplitude modulation or frequency/phase modulation of a carrier by noise vectors V_{AM} or V_{PM}, Fig. 7.35.

The carrier is a pure sinusoid of fixed amplitude, frequency and phase, whereas the noise vectors are of much smaller amplitude and their magnitude and phase vary randomly. The phase modulation derives from the *low frequency noise* of the oscillator's active device, which has the following frequency dependence. At low frequencies, the noise vector power is inversely proportional to the frequency and, close to DC, the noise decreases proportional to $1/f^3$, which is known as frequency modulated flicker noise. Then, it decreases proportional to $1/f^2$, due to frequency modulated (FM) white noise, which continues to decrease and crosses frequency f_1, denominated "corner frequency," where AM noise level is greater than the FM noise level and is constant over frequency. The corner frequency is of the order of kHz and above that frequency the FM noise is negligible compared to the AM noise.

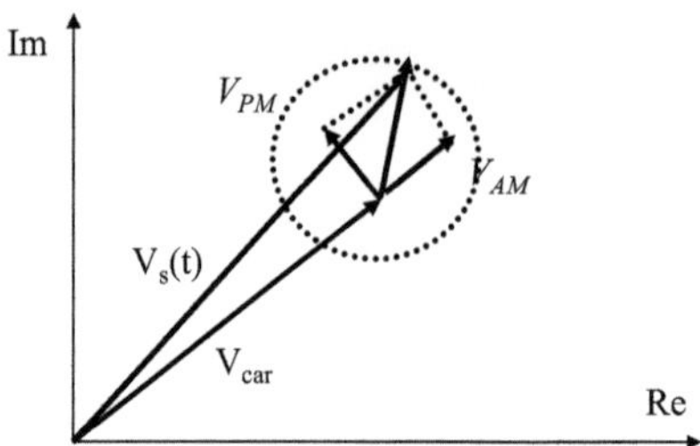

Fig. 7.35 Amplitude and phase modulation of a carrier

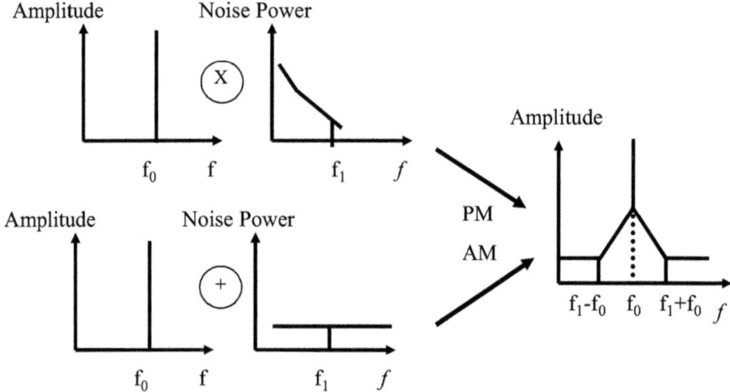

Fig. 7.36 Phase noise as a result of up-conversion

Therefore, the oscillator phase noise can be explained as a result of the nonlinear nature of oscillators, which up-converts to sideband frequencies close to the carrier. This effect is illustrated below, where AM noise is not multiplied, but added to the oscillator spectrum, Fig. 7.36.

7.9.3.1   Mathematical Treatment

Assume,

$$V_s(t) = V_c\left[1 + m(t)\right]\cos\left[2\pi f_c t + \phi_p(\tau)\right] \tag{7.23}$$

$m(t)$ represents amplitude modulation, and $\phi_p(t)$ represents phase modulation. Then, neglecting the amplitude modulation, the frequency modulation of a signal carrier, f_c, by a sinusoidal frequency f_m, is defined by,

$$f(t) = 2\pi f_c + 2\pi \Delta f_{peak} \cos(2\pi f_m t) \tag{7.24}$$

and the frequency-phase relation is given by the integration in time,

$$\phi_P(t) = \int 2\pi f(t)dt \tag{7.25}$$

Performing the integration and inserting in (7.24) into (7.23), we have,

$$V_s(t) = V_c \cos\left[2\pi f_c t + \left(\frac{\Delta f_{peak}}{f_m}\right)\sin\left(2\pi f_m t\right)\right] \tag{7.26}$$

With f_c denoting the signal carrier frequency, f_m denoting the modulating frequency or offset frequency from carrier, $V_s(t)$ denotes the amplitude of the signal, $f_p(t)$ denotes the instantaneous angle modulation, Δf_{peak} denotes the maximum frequency deviation, $\Delta f_{peak}/f_m = \beta$ denotes the frequency modulation index. Then, (7.26) may be written in the form,

$$V_s(t) = V_c \cos(2\pi f_c t)\cos\left[\beta\sin(2\pi f_m t)\right] - V_c \sin(2\pi f_c t)\sin\left[\beta\sin(2\pi f_m t)\right] \tag{7.27}$$

Now, the $\cos(\beta\sin(2\pi f_m t))$ and $\sin(\beta\cos(2\pi f_m t))$ functions may be expanded as Fourier series whose coefficients are ordinary Bessel functions of the first kind with arguments β. These functions are described by,

$$\cos\left[\beta\sin(2\pi f_m t)\right] = J_0(\beta) + \sum_{n=even} 2J_n(\beta)\cos(n2\pi f_m t) \tag{7.28}$$

$$\sin\left[\beta\sin(2\pi f_m t)\right] = \sum_{n=odd} 2J_n(\beta)\sin(n2\pi f_m t) \tag{7.29}$$

Substituting (7.28) and (7.29) into (7.27), and expanding the products of sines and cosines results in,

$$
\begin{aligned}
V_s(t) = {} & V_c J_0(\beta)\cos(2\pi f_c t) \\
& + \sum_{n=odd} V_c J_n\left[(\beta)\cos(2\pi(f_c + nf_m)t) - \cos(2\pi(f_c - nf_m)t)\right] \\
& + \sum_{n=even} V_c J_n\left[(\beta)\cos(2\pi(f_c + nf_m)t) + \cos(2\pi(f_c - nf_m))t\right]
\end{aligned}
\tag{7.30}
$$

The phase noise corresponds to a carrier frequency modulated by a single tone frequency, f_m, with a very low modulation index, $\beta \ll 1$. Therefore, only the first two terms of (7.30) need to be considered, so that the resulting spectrum consists of the carrier and two line spectra located at frequencies $(f_c + f_m)$ and $(f_c - f_m)$ shown in Fig. 7.37.

Under these conditions, it can be shown that the following approximations can be made,

$$J_0(\beta) \cong 1 \tag{7.31}$$

$$J_1(\beta) \cong 0.5\beta = 0.5(\Delta f_{peak}/f_m) \tag{7.32}$$

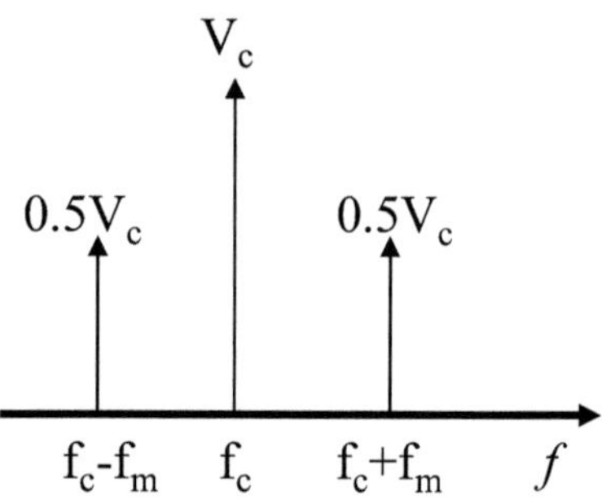

Fig. 7.37 Phase noise spectrum

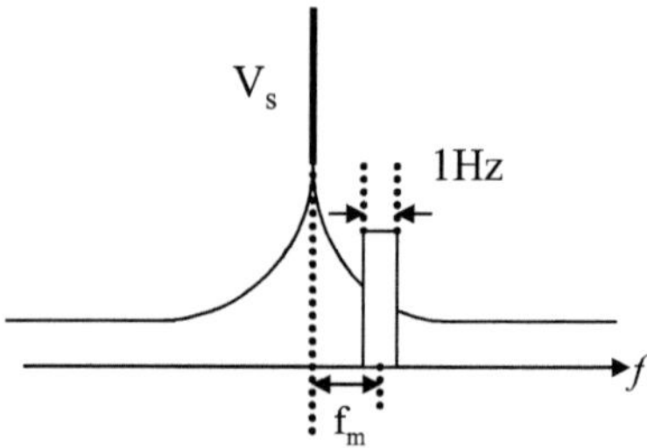

Fig. 7.38 Phase noise power

The single side band phase noise to carrier ratio is obtained by dividing the up-converted voltage at either $(f_c + f_m)$ or $(f_c - f_m)$ by the carrier voltage as follows,

$$\frac{V_{SSB}}{V_s} = 0.5\left(\frac{\Delta f_{peak}}{f_m}\right) \tag{7.33}$$

where V_{SSB} corresponds to the amplitude of the single-sideband noise. Extending the frequency components of the modulating signal from zero frequency up to the f_m frequency, the modulating signal becomes noise, whose power density in decibels is expressed as follows,

$$L(f_m) = 10\log\left(\frac{V_{SSB}}{V_s}\right)^2 = 20\log 0.5\left(\frac{\Delta f_{peak}}{f_m}\right)$$

$$= -6dB + 20\log\left(\frac{\Delta f_{peak}}{f_m}\right) \tag{7.34}$$

The phase noise power at a distance fm from the carrier is depicted in Fig. 7.38, and is measured in a 1 Hz bandwidth.

7.9.3.2 Phase Noise Mechanisms: Phenomenology

Oscillator phase noise is generated mainly by two mechanisms, distinguished by the path into which the noise is injected, namely, phase noise in the *signal path* and phase noise in the *control path* [5].

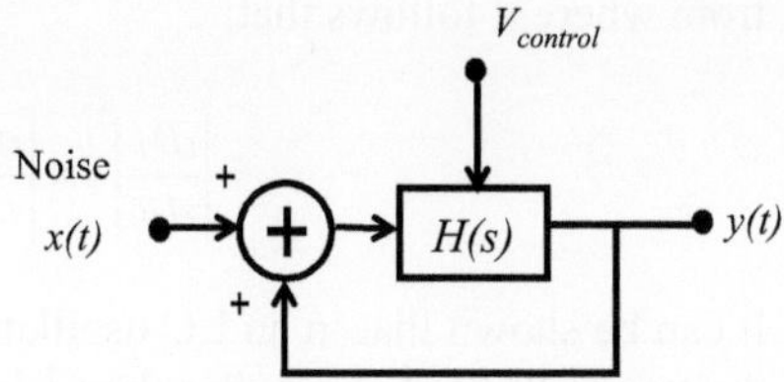

Fig. 7.39 Phase noise in signal path.

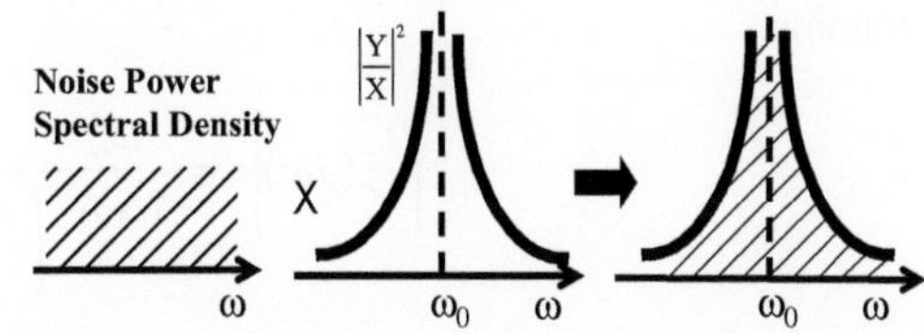

Fig. 7.40 Noise shaping in oscillators

In the signal path mechanism, noise injected into the signal path mixes with the carrier, Fig. 7.39. Representing the open loop circuit by a linear transfer function H(s), we can write,

$$\frac{Y(s)}{X(s)} = \frac{H(s)}{1 - H(s)} \qquad (7.35)$$

In the vicinity of the frequency of oscillation, $\omega = \omega_0 + \Delta\omega$, we can approximate $H(j\omega)$ with the first two terms in its Taylor expansion,

$$H(j\omega) \sim H(j\omega_0) + \Delta\omega \frac{dH}{d\omega} \qquad (7.36)$$

Because $H(j\omega_0) = +1$ and usually, $|\omega dH/d\omega << 1|$, we have,

$$\frac{Y}{X}(\omega_0 + \Delta\omega) \sim \frac{-1}{\Delta\omega \dfrac{dH}{d\omega}} \qquad (7.37)$$

This implies that a noise component at $\omega = \omega_0 + \Delta\omega$ is multiplied by $-\left(\Delta\omega dH/d\omega\right)^{-1}$ when it appears at the oscillator output. This means that the noise spectrum, Fig. 7.40, is shaped by,

$$\left|\frac{Y}{X} j(\omega_0 + \Delta\omega)^2\right| = \frac{1}{(\Delta\omega)^2 \left|\dfrac{dH}{d\omega}\right|^2} \qquad (7.38)$$

If the term $H(\omega)$ is expressed in polar form, i.e., $H(\omega) = |H|e^{j\phi}$, one has,

$$\frac{dH}{d\omega} = \left(\frac{d|H|}{d\omega} + j|H|\frac{d\phi}{d\omega}\right)e^{j\phi} \qquad (7.39)$$

from where it follows that,

$$\left|\frac{dH}{d\omega}\right|^2 = \left|\frac{d|H|}{d\omega}\right|^2 + \left|\frac{d\phi}{d\omega}\right|^2 |H|^2 \tag{7.40}$$

It can be shown that in an LC oscillator, in the vicinity of the resonance frequency, the term $|dH/d\omega|^2$ is smaller than $|d\varphi/d\omega|^2$ [5]. Since the value of $|H|$ is close to unity for steady oscillations, one can write $|dH/d\omega|^2 \sim |d\varphi/d\omega|^2$, from where one obtains,

$$\left|\frac{Y}{X}(j\omega)\right|^2 = \frac{\omega_0^2}{4\Delta\omega^2}\frac{1}{\left(\dfrac{\omega_0}{2}\right)\left(\dfrac{d\phi}{d\omega}\right)^2} \tag{7.41}$$

Now, since the Q may be expressed as,

$$Q = \frac{\omega_0}{2}\left|\frac{d\phi}{d\omega}\right| \tag{7.42}$$

we have, substituting this into (7.41),

$$\left|\frac{Y}{X}(j\omega)\right|^2 = \frac{1}{4Q^2}\left(\frac{\omega_0}{\Delta\omega}\right)^2 \tag{7.43}$$

This is the so-called *Leeson's equation* [10] It reveals the dependence of the oscillator output upon the Q of the resonator, the center frequency, and the offset frequency. This equation includes the effect of noise on both the amplitude and the phase of the carrier.

When noise is injected into the control path, Fig. 7.40a, it affects the frequency by changing the physical properties of the oscillator. For example, if a varactor is used to tune the VCO, Fig. 7.41, noise on the DC voltage applied across the diode varies the resonator capacitance and hence the resonance frequency [5].

7.9.3.3   Quantification of FM noise mechanism

The noise per unit BW is represented as a sinusoid with the same average power: $V_m cos\omega_m t$, Fig. 7.42 [5].

Denoting the VCO gain as K_{VCO} and using the narrowband FM approximation, we have,

$$v_{out}(t) \approx A_0 \cos\omega_0 t + \frac{A_0 V_m K_{VCO}}{2\omega_m}\left[\cos(\omega_0 + \omega_m)t - \cos(\omega_0 - \omega_m)t\right] \tag{7.44}$$

Thus, the noise power at $\omega_0 \pm \omega_m$ with respect to the carrier power is,

$$\left(\frac{K_{VCO}}{\omega_m}\right)^2 \frac{V_m^2}{4} \tag{7.45}$$

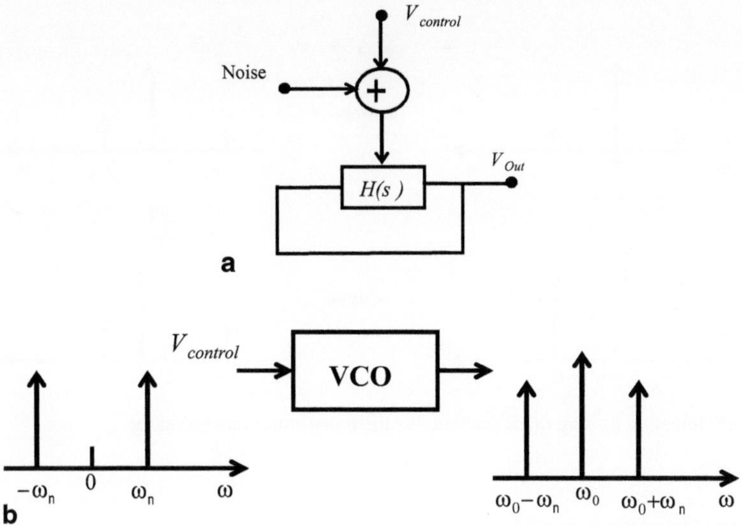

Fig. 7.41 a Phase noise in control path. **b** Modulation of VCO frequency by noise in control line

Fig. 7.42 Approximation of noise with sinusoid

FM noise arises from any source that can vary the frequency of oscillation, NOT just from the VCO control path described!

7.9.4   Oscillator Pulling and Pushing

So far, it has been assumed implicitly that the magnitude of the noise injected into the signal path is much less than that of the carrier, and this led to the shaping function for oscillators. What happens if the injected component is close to the carrier frequency and has a comparable magnitude? This is the subject of this section.

As the magnitude of the noise increases, the carrier frequency may *shift* toward the noise frequency ω_n and eventually "*lock*" to that frequency, Fig. 7.43 [5, 6]!

7.9.5   Oscillator Pulling

When desired received signal is accompanied by large interferer, if the interferer frequency is close to the LO frequency, then coupling through the mixer may pull

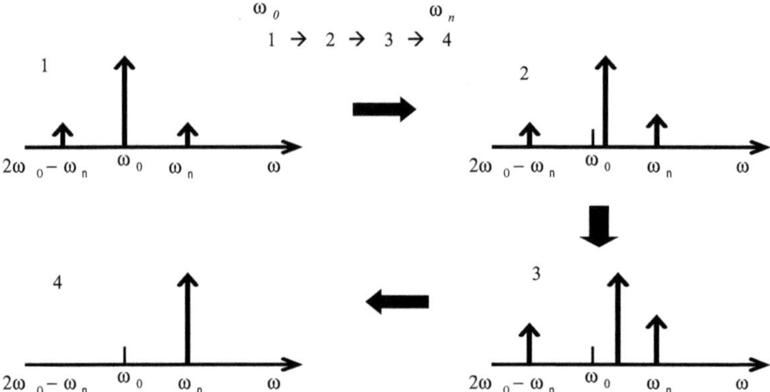

Fig. 7.43 Injection pulling of an oscillator as the noise amplitude increases

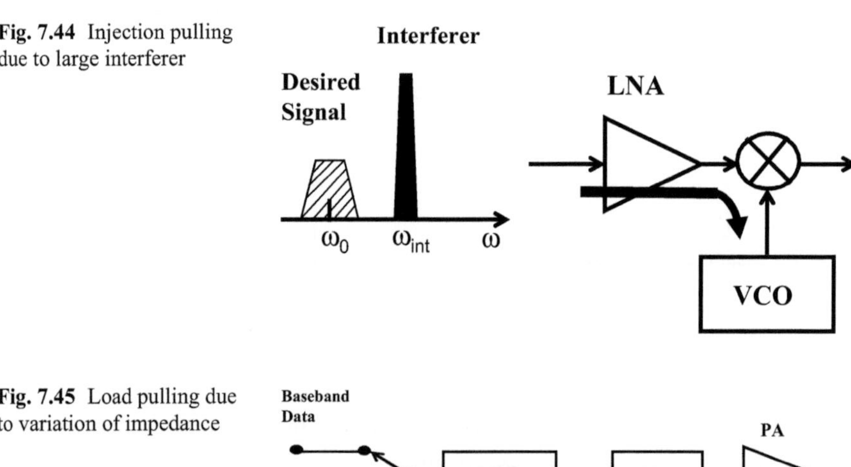

Fig. 7.44 Injection pulling due to large interferer

Fig. 7.45 Load pulling due to variation of impedance

ω_{LO} toward ω_{int}, Fig. 7.44 [4, 5]. To combat this, the VCO must be followed by a buffer stage with high reverse isolation.

Oscillator pulling may also manifest itself when there are changes in the load, Fig. 7.45.

In a GFSK modulator, for example, the VCO is first placed in feedback loop to stabilize its output frequency, then, the control voltage of the VCO is switched to the baseband, allowing the VCO to function as frequency modulator [4, 5, 7]. However, the open loop operation makes the VCO sensitive to variations in load impedance (i.e., "load pulling"). In particular, as the PA turns on and off periodically, to save energy, the VCO frequency changes considerably, demanding a high-isolation stage between the VCO and the PA [5] to avoid pulling.

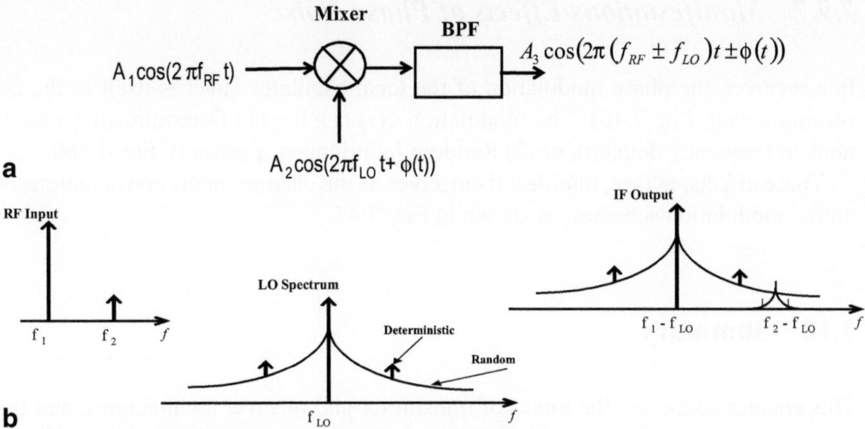

Fig. 7.46 **a** Phase noise of a Local Oscillator. **b** Spectrum

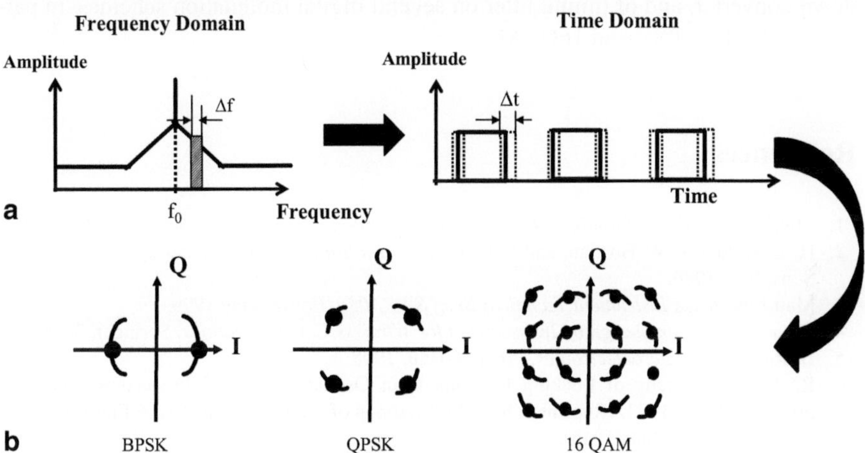

Fig. 7.47 **a** Phase noise in frequency domain spectrum and corresponding effect on modulated signal in time domain. **b** Jitter in a demodulated digital signal

7.9.6   Oscillator Pushing

VCO pushing manifests itself when, for instance, as a result of variations in the power supply, so also varies the reverse voltage across the varactor setting the frequency of oscillation. This phenomenon is referred to as "supply pushing," and is of importance in portable transceivers when the PA turns on and off [5]. It is caused by the battery's nonzero impedance, which may result in voltage variations of several hundred millivolts [5].

7.9.7　Manifestations/Effects of Phase Noise

In a receiver, the phase modulation of the local oscillator attaches itself to the incoming signal, Fig. 7.46a. The modulation $\phi(t)$ can be: (1) Deterministic (δ functions in frequency domain), or (2) Random (continuous spectrum), Fig. 7.46b.

These effects, in turn, manifest themselves as the changes in the constellations of digital modulation schemes, as shown in Fig. 7.47.

7.10　Summary

This chapter addressed the topics of transmitter and receiver architectures, and the performance of oscillators. In particular, the heterodyne and homodyne architectures were discussed, together with the image-reject, digital-IF, and sub-sampling architectures. The phase noise performance of oscillators was then discussed, followed by the impacts of oscillator pulling and pushing. The chapter concluded with a pictorial presentation of the manifestations of phase noise on the spectrum of a down converter, and of timing jitter on several digital modulation schemes, in particular, BPSK, QPSK and 16QAM.

References

1. Mischa Schwartz, *Information Transmission, Modulation, and Noise*, McGraw-Hill, 1970.
2. H. L. Krauss, C.W. Bostian, and F. H. Raab, *Solid State Radio Engineering*, John Wiley & Sons, Inc., 1980.
3. Matt Loy, *Texas Instrument Technical Brief SWRA030*, Editor, May 1999.
4. Q. Gu, *RF System Design of Transceivers for Wireless Communications*, Springer, 2005.
5. B. Razavi, *RF Microelectronics*, Prentice-Hall, 1998.
6. R. Adler, "A Study of Locking Phenomena in Oscillators," IEEE Proceedings, vol. 61, pp. 1380–1385, 1973. (Reprinted from Proceedings of the Institute of Radio Engineers, vol. 34, pp. 351–357, June 1946).
7. B. Razavi, "RF Transmitter Architectures and Circuits," 1999 *Proc. IEEE Custom Integrated Circuits Conf.*, pp. 197–204.
8. P.-I. Mak, U. Seng-Pan, and R. P. Martins, "Transceiver Architecture Selection: Review, State-of-the-Art Survey, and Case Study," *IEEE Circuits and Systems Magazine*, Second Quarter 2007, pp. 6–25.
9. R. M. Cerda, "Impact of ultralow phase noise oscillators on system performance," *RF Design*, July 2006, pp. 28–34.
10. D. B. Leeson, "A Simple Model of Feedback Oscillator Noise Spectrum", *Proc. IEEE* Vol. 54, No. 2, pp. 329–330, 1966.

Chapter 8
Case Studies

Abstract In this chapter, we integrate the knowledge presented in all the previous chapters. Its aim is to discuss and clarify in a tutorial fashion two papers that expose the engineering considerations and systems engineering thought processes behind the design of real-life wireless systems, in particular, a WCDMA receiver and a Long Term Evolution (LTE) receiver.

8.1 Introduction

In this chapter, we integrate the knowledge presented in all the previous chapters. Its aim is to discuss and clarify in a tutorial fashion two papers [2] and [8], that expose the engineering considerations and systems engineering thought processes behind the design of real-life wireless systems, in particular, pertaining to receivers operating as per the WCDMA [1, 2] and a Long Term Evolution (LTE) [8] "standards"[1], see the Appendix.

8.2 WCDMA Receiver

The WCDMA standard has been developed for third-generation (3G) mobile systems [3]. This standard was developed to offer enhanced data rates and capacity over second-generation systems. WCDMA employs frequency-division duplexing (FDD), where the transmit (Tx) signal occupies the 1920–1980-MHz band, while the receive (Rx) signal occupies the 2110–2170-MHz band. Thus, WCDMA is a full-duplex system in which the handset transmits and receives simultaneously in these separate Tx and Rx bands.

Many system performance specifications derive from this full-duplex feature [3]. WCDMA mobile receivers require both very high sensitivity and very high

---

[1] The implementation of wireless connectivity relies upon the definition of so-called *wireless standards*. Each standard embodies the precise set of parameters that dictate the architecture and software design of wireless systems operating under the given standard, to effect compatible communication with other systems, also operating within the standard.

H. J. De Los Santos et al., *Radio Systems Engineering,*
DOI 10.1007/978-3-319-07326-2_8, © The Authors 2015

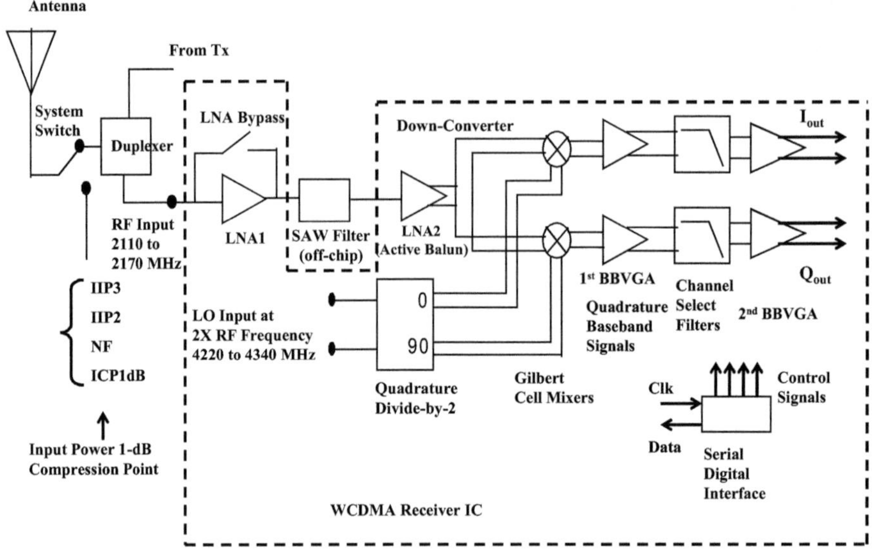

Fig. 8.1 Overall receiver block diagram. Specifications are: IIP3, IIP2, NF, and 1CP1dB

linearity, with stringent out-of-band linearity requirements due to the presence of transmit leakage. Specification for key analog/RF and digital performance requirements are given in the 3GPP Technical Specification document [3]; the reader is urged to thoroughly examine the 3GPP document [3] before proceeding.

8.3 WCDMA Receiver Analog Systems Requirements

The overall system to be analyzed is shown in Fig. 8.1. Analog RF system requirements such as noise figure and linearity for the IC design are defined based on the 3GPP specification, combined with a set of assumed operating parameters including maximum transmitter leakage seen at the low-noise amplifier (LNA) and maximum front-end duplexer loss.

The WCDMA receiver architecture is that of a direct-conversion receiver [1, 2]. Some of the advantages of this architecture are as follows: (1) Elimination of one or more IF filters; (2) The need for only one frequency synthesizer; (3) A cleaner frequency plan with fewer unwanted spurious mixer outputs; and (4) The potential to accommodate different channel bandwidths in a straightforward way. However, the transmit leakage in WCDMA makes out-of-band linearity requirements difficult to meet and strongly influences the circuit designs [2].

With respect to Fig. 8.2, the operation of the receiver is as follows (the numbers refer to points in the schematic drawing). The received WCDMA signal from the antenna **1** passes through a system switch **2**, a duplexer **3**, and into a switched-gain LNA, LNA1 (15-dB gain or 4-dB loss) **4**. The output of LNA1 goes off-chip to a bandpass surface acoustic wave (SAW) filter, and then comes back on-chip to a

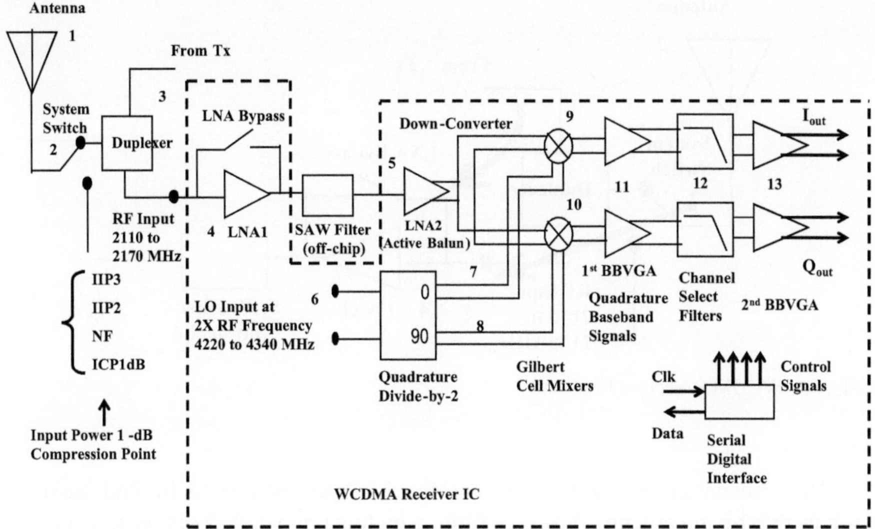

Fig. 8.2 Overall receiver block diagram

mixer preamplifier, LNA2 (12-dB gain) **5**. LNA2 provides differential RF signals to the double-balanced Gilbert mixers, each of which has 6 dB of gain. The local oscillator (LO) signal **6** comes onto the chip differentially at twice the incoming RF frequency and is then frequency-divided by two to provide differential quadrature LOs **7** and **8** to the mixers **9** and **10**; this approach minimizes LO leakage through the package to the inputs of LNA1 and LNA2 [1, 2]. The mixer outputs feed the first baseband variable-gain amplifiers (BBVGA1) **11**, which provide five selectable gain settings from 16 to -8 dB in 6-dB steps. BBVGA1 is followed by a five-pole, two-zero transconductance–capacitance (Gm-C) filter **12** with 6 dB of gain and a 3-dB cutoff frequency of 1.92 MHz. A second BBVGA **13** provides up to 50 dB of gain in 1-dB steps, bringing the baseband signal up to 62.5 mV rms (typical). BBVGA2 outputs go off-chip to 6-b analog-to-digital converters (ADCs) located on the digital baseband IC [1, 2].

8.4   WCDMA System Performance Requirements

The system-level performance requirements for the receiver IC, Fig. 8.3, are derived from the WCDMA RF specifications in the 3GPP document [3], combined with the assumption of a maximum level of the leakage power from the transmitter reaching the LNA input.

The noise Fig. (NF) requirements for the receiver IC, in turn, are derived directly from the WCDMA RF sensitivity test, together with the assumption that the maximum of signal loss from the antenna to the LNA (through the antenna switch and the diplexer filter **2**) is 4 dB. A maximum of 5 dB LNA-referred NF can be tolerated, while providing approximately 2 dB of margin for other system degradations.

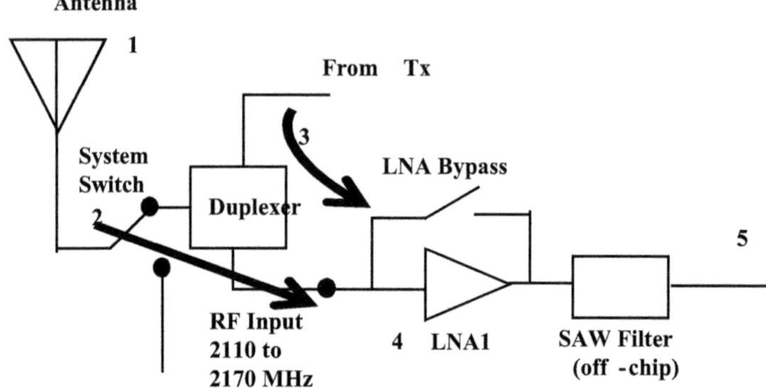

Fig. 8.3 WCDMA receiver front-end

The transmitter power that reaches the LNA arises due to limited isolation achievable by the antenna duplexer filter, which is typically 50 dB. With a Tx output **3** power of $+28$ dBm, the leakage level present at the receiver LNA input **4** is -22 dBm. This strong level of leakage can combine with out-of-band blocker signals to produce in-channel third-order intermodulation (IM3) products which *desensitize* the receiver. The blocking level specifications for the receiver are given in Table 8.1 [2].

Due to the fact that the WCDMA specification stipulates four blocking bands, four out-of-band linearity requirements are derived by allocating approximately 33 % of the signal distortion budget for each IM3 term; this allows margin for other receiver system degradations. The in-band IIP3 requirement is derived directly from the WCDMA in-band IM RF test condition with a signal distortion allocation of approximately 16 % for the IM3 product to leave margin for other degradation such as blocker leakage and second-order intermodulation distortion (IM2) distortion [1, 2].

The origin of the out-of-band third-order linearity requirements stems from the transmitter leakage reaching the LNA and mixing with blocker signals below the receiver band; this produces third-order intermodulation (IM3) distortion in the receiver channel passband [2]. A summary of the frequency bands relevant to the transmit/blocker IM3 distortion problem is given in Table 8.1. An estimate is given, in each band, of the minimum amount of insertion loss provided by either the duplexer alone (for transmitter leakage from the duplexer transmit port **3**, Fig. 8.3, to the receive port **4**), or the combined switch/duplexer (for out-of-band blockers from the antenna to the duplexer receive port). By cascading the blocker power levels at the antenna with minimum switch and duplexer attenuation, one can determine the maximum blocker power level at the LNA input, so that the band-specific LNA-referred third-order linearity and input compression requirements for the receiver IC can be defined [2]. The attenuation levels cited in the Table 8.1 are just typical values since, depending on the specific properties of the switch/duplexer filter design and vendor, they may vary widely [2].

At the LNA input (**4**, Fig. 8.3), the strong transmit leakage term can also deteriorate the performance of a direct-conversion receiver via the IM2 in the mixer. To

Table 8.1 Blocking levels at the LNA input

Band	Frequency Range (MHz)	Power Level at Power Amplifier output or antenna in (dBm)	Minimum Attenuation (dB)	Power level at LNA (dBm)
Transmit band	1920–1980	+29 PA	50 (Tx–Rx)	−21
Blocking band	2050–2075	−44 (antenna)	2 (ant–Rx)	−46
Blocking band 2	2025–2050	−30 (antenna)	10 (ant–Rx)	−40
Blocking band 3 (above Tx band)	2015–2025	−15 (antenna)	25 (ant–Rx)	−40
Blocking band 3 (below Tx band)	1679–1840	−15 (antenna)	44 (ant–Rx)	−59

make sure that there is negligible degradation (<0.1 dB) from the down-converter IM2 under WCDMA sensitivity test conditions, see [3], the budget allocates only 2% of the total SNR distortion budget to the transmit-band IM2 in the sensitivity test [1, 2]. This results in an LNA-referred second-order intercept point (IIP2) requirement across the transmit band of+ 72 dBm. To achieve this high IIP2 (in addition to the blocking-band IIP3), it is necessary to use an inter-stage RF bandpass (SAW) filter between the first LNA and the down-converter input, thus providing at least 25 dB of attenuation in the Tx band.

8.5   WCDMA Receiver System Operating Conditions

The key RF system performance requirements for the direct-conversion receiver include the LNA-referred noise Fig. (NF), the input-referred second- and third-order intercept points (IIP2, IIP3), and the input power 1-dB compression point (ICP1dB), Fig. 8.4.

It should be kept in mind that the front-end switch and RF filter characteristics can vary significantly, depending on vendor and design. Therefore, the LNA-referred receiver performance requirements, such as NF, IIP2, IIP3, and ICP1dB, cannot be directly derived from the WCDMA specification [2, 3]. As a result, completing the performance requirements needs that some important parameters be obtained from actual operating-conditions. These include the maximum transmitter leakage power at the LNA input port, the insertion loss and blocking characteristics of the antenna switch and duplexer filter, and the performance of the baseband digital-signal processor (DSP) demodulator [2]. The assumptions made for each of these parameters are described in the following.

8.5.1   Transmitter Leakage at LNA input

The receiver IC considered was designed for use in a WCDMA FDD mobile device [2]. FDD means that the transceiver was designed to achieve full duplex functionality

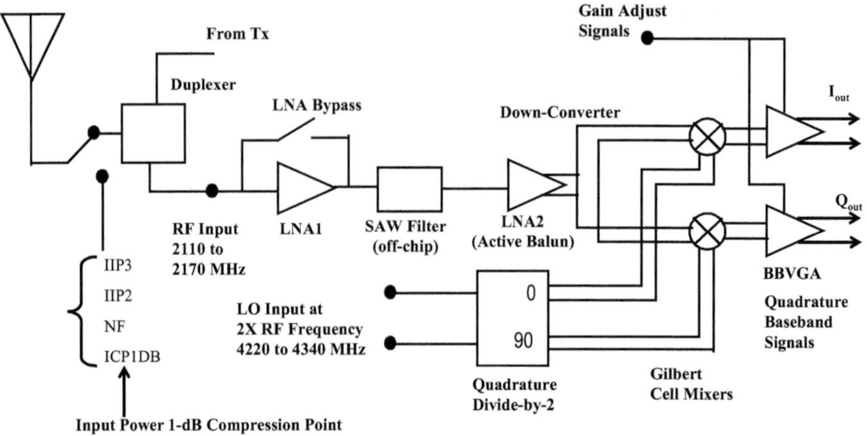

Fig. 8.4 Block diagram of the WCDMA receiver chip

by being simultaneously active at different transmit (Tx) and receive (Rx) frequencies. Clearly, because of this there is a leakage from the transmitter that reaches the LNA input. Figure 8.3.

To implement an FDD system it is essential for the radio to use a three-port duplexer filter so as to isolate the relatively high-power transmit signal from a small received signal at the antenna. Since the duplexer isolation is limited, some amount of residual transmitter leakage power appears at the receiver LNA input. Therefore, for the system design considered in this paper, a maximum operating transmitter leakage of -21 dBm at the LNA input was assumed [2].

This level stems from the assumption of $+29$ dBm at the transmit power-amplifier output (or duplexer transmit port) to achieve $+27$ dBm class-2 (see [3]) transmit power at the antenna output (the combined transmit duplexer filter and antenna switch loss is assumed to be 2 dB), and an assumed 50 dB of isolation from the duplexer transmit output to receive input port (**3** to **4** in Fig. 8.3.) This leakage level is one of the key parameters driving the compression and out-of-band linearity requirements of the receiver. **Recall the feedthrough from Tx path to Rx path discussion in Chap. 7.**

8.5.2 *Duplexer Filter Frequency Response*

The out-of-band third-order linearity requirements of the receiver stem from the transmitter leakage arriving at the LNA and mixing with blocker signals below the receiver band, which produce third-order intermodulation (IM3) distortion in the passband of the receiver channel. The blocking bands defined in the WCDMA specification, Table 8.1, correspond to blocker power levels that must be tolerated by the receiver [3]. Refer again to Table 8.1 for a summary of the frequency bands relevant to the transmit/blocker IM3 distortion problem.

8.5.3   *Required Es/N$_0$ and Receiver Implementation Loss*

To assess compliance with the specifications, the RF tests of a WCDMA receiver use coded bit-error rate (BER) as the performance metric [2]. Passing these tests [1] requires that a coded BER of 0.1 % be met under a range of test conditions designed to stress sensitivity, linearity, and selectivity of the receiver. From reference simulation results, the 12.2-Kb/s modulation used for the RF tests requires approximately 0.9 dB Es/N_0 (channel-symbol energy-to-noise power spectral density ratio) or de-spread channel symbol signal-to-noise ratio (SNR) to achieve 0.1 % BER [1, 2]. The 0.9-dB Es/No is increased by a receiver *implementation-loss*[2] factor to allow margin for practical effects such as analog filter amplitude or phase distortion in the receiver and non-ideal matched filtering or imperfect channel estimation in the baseband DSP [1, 2]. An implementation loss of less than 1 dB is attainable in a high-performance radio and DSP combination, however, to be conservative, an implementation loss of 2 dB was chosen to build margin into the system specification requirements. Therefore, with this loss, an Es/No of 2.9 dB must be maintained in order to pass the BER requirement for the RF tests [1, 2].

8.5.4   *Receiver Noise*

The noise performance of the receiver was characterized by its input-referred noise figure [2]. The noise figure was measured by integrating the noise power across the full received signal bandwidth of 3.84 MHz [2]. The total integrated noise, as opposed to the spot noise, is needed in order to accurately predict the achievable system sensitivity. Because the receiver passband may have significant gain ripple in the cascaded LNA→bandpass→filter(BPF)→LNA2→Mixer transfer function, Fig. 8.3, the noise must be characterized at each of the 12 center frequencies across the UMTS Terrestrial Radio Access (UTRA) receive band (2110–2170 MHz) [3].

The noise figure requirement of the receiver is driven by: (1) The received power level, as specified in the WCDMA sensitivity test (-117 dBm code channel at the antenna); and (2) the required channel Es/N_0 for the operation at 0.1 % BER [2]. An LNA-referred noise figure of less than 5 dB was targeted to provide ~2 dB of sensitivity margin, with a 4 dB loss in the front-end RF switch/filter. This result can be derived using the code channel spreading gain for the 12.2-Kb/s WCDMA modulation coupled with the thermal noise power of -108.2 dBm in the channel BW of 3.84 MHz [1, 2].

WCDMA uses Direct Sequence spreading, see Chap. 2, [4], where the spreading process is done by directly combining the baseband information to a high chip rate binary code. The Spreading Factor is the ratio of the chips (UMTS = 3.84 Mchips/s) to baseband information rate. Spreading factors vary from 4 to 512 in FDD UMTS;

---

[2] **Implementation Loss:** Is the difference between theoretical performance and actual performance due to implementation. It arises from timing errors, frequency offset, finite rise-and fall times of the waveforms, etc

the spreading process gain can in expressed in dB (Spreading factor 128=21 dB gain) [2]. The spreading gain effectively lowers the thermal noise power to a level of -129.2 dBm because the code length of 128 used with the 12.2 Kb/s modulation reduces the thermal noise power by a factor of 1/128, or -21 dB [2]. With a noise figure of 5 dB, the thermal noise power is increased to -124.2 dBm.

The maximum antenna-to-LNA loss of 4 dB results in a LNA-referred signal power of -121 dBm at sensitivity. The de-spread [4] channel-symbol SNR that results was 3.2 dB, which provides 2.3 dB of margin beyond the 0.9-dB minimum requirement for operation at 0.1 % BER [2].

8.5.5   Third-Order Linearity

The input-referred third-order intercept point (IIP3) performance of a receiver characterizes its third-order linearity [2]. As discussed in Chap. 6, in a receiver the third-order nonlinearity manifests itself as a distortion product that appears in-channel, caused by two interferers offset by Δf and $2\Delta f$ from the desired channel in frequency. The interference signals occur both within the receiver passband (in-band IM3 interference) and outside the receiver passband (out-of-band IM3 interference [2].

To determine the in-band IM3 requirement, an explicit IM test was performed on the receiver under the conditions of the WCDMA specifications [10 MHz offset continuous-wave (CW) tone, 20 MHz offset WCDMA interferer, both at a power level of -46 dBm at the antenna] [3]. However, the out-of-band IM3 requirement is defined only implicitly by the need of the FDD receiver to pass the out-of-band blocking desensitization tests in the presence of a large transmit leakage signal at the antenna [2].

From the WCDMA specifications, there are four blocking regions (summarized in Table 8.1) that can produce in-channel IM3 products when mixed with the transmitter leakage signal. Table 8.2 shows the targeted minimum IIP3 for the in-band and the four out-of-band scenarios [3].

The targeted in-band IIP3 is greater than 19.5 dBm, while the targeted out-of-band IIP3 is in the range of 4.5–1.5 dBm, depending on the blocking band under consideration.

8.5.6   Second-Order Linearity

The second-order linearity performance of a receiver is characterized in terms of its input-referred second-order intercept point (IIP2) for interference signals that lie within both in-band and out-of-band [2] frequencies. As seen previously, a nonlinearity of second-order in an amplifier produces distortion products that lie in the baseband which, in turn, can add time-varying DC offsets and in-channel distortion to the down-converted signal [2]. In the case of WCDMA, the solution to the DC offset problem involves using a high-pass filter to remove the bottom 1–5 kHz

Table 8.2 IIP3 System Performance Results (Measured Worst-Case Figures) [2]

Parameter	Unit	Target	Measured	Description
IIP3 (in-band)	dBm	>−19.5	−18.6	In-band IM3 m 10 MHz tone spacing
IIP3 (Band1, Tx)	dBm	>−4.7	−0.9	Band1/Tx IM3
IIP3 (Band2, Tx)	dBm	>+1.3	0.1	Band2/Tx IM3
IIP3 (Band3, Tx)	dBm	>+1.3	0.8	Band3 above Tx/Tx IM3
IIP3 (Band3, Tx)	dBm	>+0.8	1.3	Tx/Band3 below Tx IM3

Table 8.3 IIP2 System Performance Results (Measured Worst-Case Figures) [2]

Parameter	Unit	Target	Measured	Description
IIP2 (in-band)	dBm	>+10	29.4	In-band IM2 tested at +/-15 MHz off-set, 1 MHz tone spacing
IIP3 (Tx-Band)	dBm	>+72	-0.9	Tx-Band IM2, 1 MHz tone spacing

portion of the signal bandwidth. However, since it is not possible to filter away distortion originating at higher frequencies that fall in-band, it is necessary to constraint the maximum allowable second-order nonlinearity as a function of blocker frequency. The IIP2 requirements are listed in Table 8.3.

As indicated in Table 8.3, a requirement for in-band IIP2 of $+10$ dBm or greater stems from the maximum in-band blocker that the receiver must tolerate. This RF specification involves, for WCDMA, a blocker level at a 15 MHz offset from the desired received channel, with a power level of -44 dBm at the antenna.

On the other hand, an out-of-band (transmit) IIP2 requirement of $+72$ dBm or greater is determined by the large transmit leakage signal $(-21$ dBm) at the LNA input. In order to meet the large IIP2 requirement of the down-converter at the transmit offset frequency, without imposing unrealizable constraints on the IIP2 of the down-converter, one must ensure that the inter-stage LNA/LNA2 RF bandpass filter has sufficient transmit-path-to-receive-path *isolation* (a minimum level of approximately 25 dB is desired for the system design described here) [2].

8.5.7   *Input 1-dB Compression*

The input 1 dB compression points are specified for frequency signals processed by both in-band (high- and low-LNA RF gain) and transmit-band (high-LNA RF gain only) circuits [2]. The high-gain in-band compression requirement is derived from the in-band IM test condition, which, with the receiver in high-gain mode, produces a composite -43-dBm signal level at the antenna input. To provide enough margin over the average power level of the received WCDMA interferer signal, a 10 dB margin is added to the average input power level; this results in a desired antenna-referred in-band compression point of -33 dBm [2]. This, together with a minimum switch/duplexer insertion loss of 2 dB results in a high-gain in-band 1 dB compression point of -35 dBm or better, referred to the LNA input of the receiver. However, since the maximum WCDMA input level test places a -25 dBm signal at

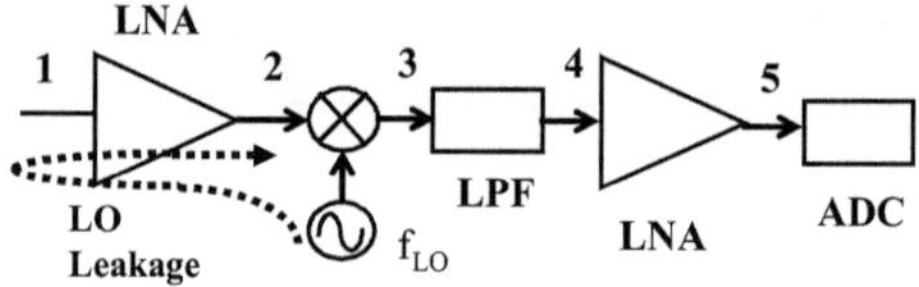

Fig. 8.5 Source of DC-offsets in direct-conversion receiver

the LNA input, a low-gain in-band compression point of -17 dBm is chosen. This low-gain compression specification does not place a more stringent requirement on the LNA2/mixer circuitry than the high-gain requirement because the LNA is switched into a low-gain state in this mode, which lowers the LNA-to-LNA2 voltage gain by about 18 dB.

The specification for compression in the transmit-band is derived from the maximum assumed transmitter leakage power experienced at the LNA input. For instance, a leakage of -21 dBm with 7 dB margin added to accommodate the WCDMA mobile unit crest factor [3], results in a desired minimum ICP1dB of -14 dBm at the LNA input for signals in the 1920–1980 MHz transmit band [2]. Then, depending on the characteristics of the RF inter-stage bandpass filter, this ICP1dB requirement may be limited by either the LNA or the LNA2/mixer. Meeting this ICP1dB requirement is critical because if it is not met, the receiver may be desensitized by the transmitter leakage under conditions of small received power and high transmit power due to small-signal gain compression and/or noise figure degradation (remember, **Chap. 5**, that compression implies reduced gain, and reduced gain implies lower noise figure) in the receiver amplifiers or mixers.

8.5.8   LO–RF Leakage

As discussed in Chap. 7, a direct-conversion receiver is susceptible to desensitization due to third-order nonlinear distortion product originating from the cross-modulation of the transmitter leakage signal and the LO–RF leakage signal at the LNA input, Fig. 8.5.

Even though it is possible to, at least in theory, deal with a large LO-RF leakage by improving the input third-order linearity, this solution is not practical because a high linearity requirement would result in an unacceptably large power draw. (Linearity is improved by increasing IP3, which in turn is accomplished by increasing current x power product, i.e., power consumption/draw). Therefore, a constraint is placed on the LO-RF leakage level so that it does not increase the needed out-of-band third-order linearity beyond that level needed from the worst case Tx/blocking band linearity requirement [2]. Based on these considerations, a maximum high-LNA-gain-mode LO–RF leakage power of -80 dBm at the LNA input was chosen as a design target. On the other hand, a low-LNA-gain-mode LO–RF leakage power at the antenna of -60 dBm must also be maintained to comply with WCDMA standards [2].

Table 8.4 Summary of LTE Features

Family	Primary Use	Radio Technology	Downlink (Mb/s)	Uplink (Mb/s)	Notes
UMTS/4GSM	General 4G	OFDMA/ MIMO/ SC-FDMA	100 (in 20 MHz band-width)	50 (in 20 MHz bandwidth)	LTE-Advanced update expected to offer peak rates up to 1 Gb/s fixed speeds and 100 Mb/s mobile users

8.6   LTE (Long Term Evolution) Receiver (4G)

The 3GPP [3] Long Term Evolution (LTE) is the latest standard in the mobile network technology evolution that produced the GSM/EDGE and UMTS/HSPA network technologies (See the Appendix). In fact, is has been construed as the first step towards true 4G technologies. LTE is frequently marketed as 4G, however, the first-release LTE does not fully comply with the 4G requirements [5, 6].

A truly 4G technology should be capable of download speeds of 100 Mb/s and 1 Gb/s from moving or pedestrian points, respectively [4, 5]. To avoid consumer confusion with the terms 3.5G or 3.9G, that began appearing, the industry decided to market LTE as "4G LTE" [8]. The main advantages of LTE are as follows:

1. High throughput;
2. Low latency;
3. Plug and play;
4. FDD and TDD in the same platform;
5. An improved end-user experience;
6. A simple architecture resulting in low operating costs.

LTE is also targeting the support of seamless passing to cell towers with older network technology such as GSM, cdmaOne, UMTS, and CDMA2000. A summary of the LTE capabilities is given in Table 8.4.

LTE is the next major step in mobile radio communications. The advantages of LTE are high downlink data rate of 100 Mbit/s (50 Mbit/s in uplink), FDD and TDD in the same platform with simple direct-conversion architecture. LTE also should have backward compatibility to existing networks such as GSM, CDMA and WCDMA.

8.7   Study of LTE Receiver Front-End [8]

In this section, we integrate knowledge presented in this book by studying the reasoning behind the systems analysis of an LTE receiver [8]. A step-by-step tutorial analysis pertinent to the LTE standard and based on the *SystemVue* software tool, is presented in the Tutorials sections.

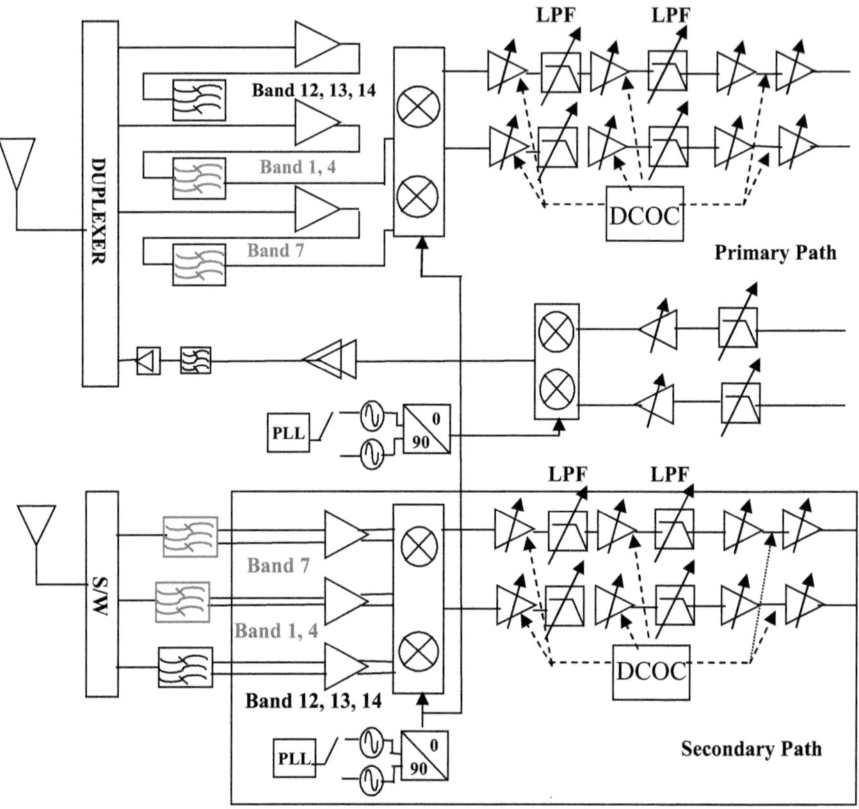

Fig. 8.6 LTE transceiver block diagram. *After* [8]

A block diagram of an LTE receiver that could support a single or multiple bands, depending on the requirements of service vendors, is shown in Fig. 8.6 [8].

The LTE standard, released by 3GPP [3], defines a multiple-in, multiple-out (MIMO) specification. The LTE transceiver of Fig 8.6, in particular, has two independent receiver chains to meet a 2X2 MIMO configuration optimized to provide a downlink in the UHF TV bands range of 728–768 MHz [8].

Since the initial LTE networks are FDD systems, with Tx and Rx on different frequencies, ensuring that the transmit energy does not leak into the receiver requires significant filtering. In particular, the Tx signal is usually large enough to saturate Rx blocks, including LNAs and mixers. As a result, in an LTE transceiver it is essential to carefully budget the Rx block linearity, so it doesn't degrade the Rx EVM and the gain, due to the strong Tx leakage [8].

The approach to reducing the Tx leakage into the Rx, entails employing an external SAW filter. This filter, which is inserted after the LNA, Fig. 8.6, provides additional leakage attenuation and, therefore, relaxes the linearity requirements of the subsequent components.

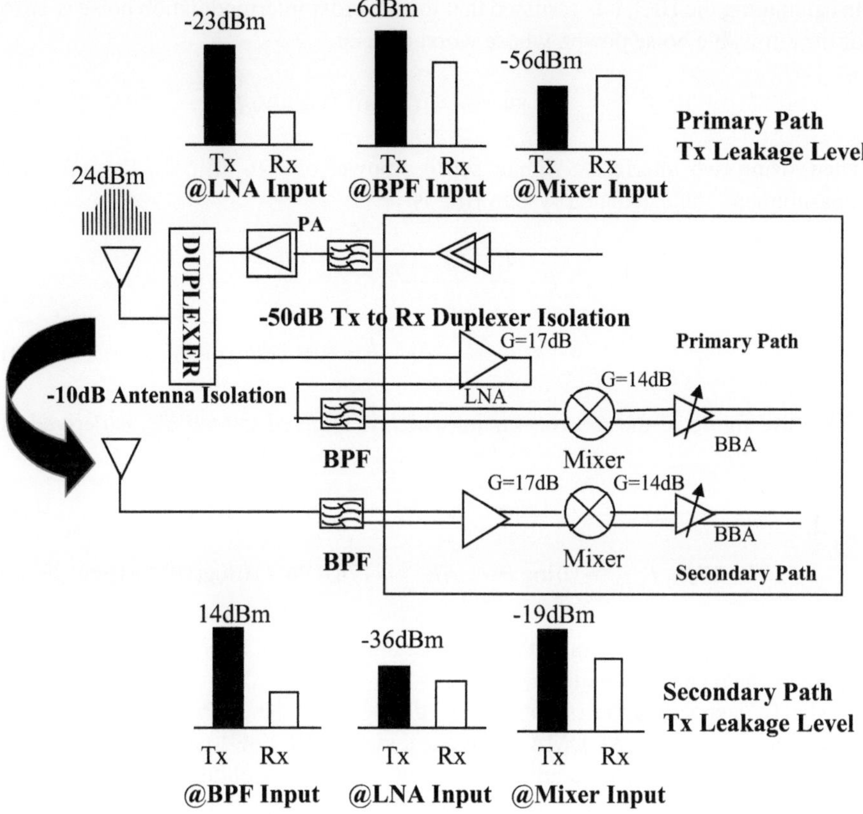

Fig. 8.7 Tx leakage level before the mixer at each receiver chain. *After* [8]

In the secondary path of Fig 8.6, an architecture that does not require a SAW filter is, in turn, chosen. In this architecture, relieving the linearity requirement of the mixer demands that the SAW filter before the LNA exhibit good selectivity. A concrete example is shown in Fig. 8.7.

Figure 8.7 shows the Tx leakage level in each signal path, before the mixer. These levels correspond to the utilization of external SAW filters having a channel selectivity of 50 dB.

In extracting the system linearity for the LTE receiver, the minimum sensitivity level of QPSK is assumed from LTE standard (3GPP TS 36.101 version 8.5.1) [3]. In particular, the required minimum sensitivity of QPSK at 10 MHz bandwidth is -94 dBm. Based on the test cases described in the blocking characteristics, section 7.6 in the 3GPP standard [3], 6 dB is added to the desired signal. If the minimum carrier-to-noise ratio is 1 dB, with wideband intermodulation, then the allowable noise power at the Rx input is -89 dBm (8.1).

$$N_i < S_i - SNR_{QPSK1/3} = -94 + 6 - (+1) = -89 dBm \tag{8.1}$$

In calculating the IIP3, it is assumed that the 3rd-order intermodulation noise is 10% of the allowable noise power, whose worst case is:

$$P_{N\,3rd\,IMD} = -89dBm + 10\log(0.1) = -99dBm \tag{8.2}$$

Then, using two interfering signals set to a power of -46 dBm for the reference measurement, the calculated system IIP3 is:

$$IIP3 \geq P_i + \frac{P_i - P_{N3rd\,IMD}}{2}$$

$$= -46 + \frac{-46 - (-99)}{2} = -19.5dBm \tag{8.3}$$

The noise figure of the receiver chain, which is targeted to be 9 dB, is derived as follows:

$$P_{in,min} = -174dBm / Hz + NF + 10\log B + CNR_{min} \tag{8.4}$$

$$NF = 174 + P_{in,min} - 10\log B - CNR_{min} = 174 - 94 - 10\log(10^6) - 1 = 9$$

Excluding the insertion loss in the duplexer and the SAW filter, the required noise figure of the receiver is 6 dB.

The IIP2 specification is not easy to derive from the 3GPP specification [3]. However, usually, for WCDMA receivers, a system IIP2 of 40 dB minimum is required by handset manufacturers [7]. In the filter-less architecture, the effect of leakage from the Tx, among the second-order intermodulation products (IM2), is greater than that from blockers or interferers. Now, it has been determined [8] that, since the IM2 distortion occurs in the down-converting mixer of a direct-conversion receiver, the bottleneck of the entire system linearity is set by the IIP2 of the mixer in the secondary path. Therefore, assuming that the IM2 product level due to Tx leakage needs to be at least 16 dB lower than the allowable noise, the maximum allowable IIP2 before the LNA due to Tx leakage can be estimated as [8, 9]:

$$P_{IM2} < N_i - 16dBm = -89 - 16 = -105dBm \tag{8.5}$$

and

$$IIP2 = 2P_i - P_{IM2} = 2 \times (-36) - 105 = 33dBm \tag{8.6}$$

Adding to this the LNA gain of 17 dB, Fig. 8.6, the mixer IIP2 must be greater than 50 dBm.

Given these results, further specifications for the LNA and mixer, are determined from link budget analysis, including filter and baseband (BBA) requirements. A software tool such as *SystemVue*, is the ideal environment for this optimization exercise. See the tutorials.

8.8 Summary

In this chapter, we integrated the knowledge presented in all the previous chapters. Its aim was to discuss and clarify in a tutorial fashion two papers, [2] and [8], that expose the engineering considerations and systems engineering thought processes behind the design of real-life wireless systems, in particular, a WCDMA receiver [1, 2] and a Long Term Evolution (LTE) receiver [8].

References

1. S. K. Reynolds, B. A. Floyd, T. J. Beukema, T. Zwick, and U. R. Pfeiffer, "Design and Compliance Testing of a SiGe WCDMA Receiver IC With Integrated Analog Baseband," *Proc. IEEE*, Vol. 93, No. 9, Sept. 2005, pp. 1624–1636.
2. S. K. Reynolds, B. A. Floyd, T. J. Beukema, T. Zwick, U. R. Pfeiffer and H. A. Ainspan,"A directconversion receiver integrated circuit for WCDMA mobile systems," *IBM J. RES. &*; *DEV.* VOL. 47 NO. 2/3 MARCH/MAY 2003, pp. 337–353.
3. 3GPP TS 25.101 V5.2.0 (2002–03) *Technical Specification* 3rd Generation Partnership Project; Technical Specification Group Radio Access Networks; UE Radio Transmission and Reception (FDD) (Release 5)
4. R. C. Dixon, *Spread Spectrum Systems with Commercial Applications*, Third Edition, John Wiley & Sons, Inc., New York, 1994.
5. Available: [Online]: http://en.wikipedia.org/wiki/LTE_(telecommunication)
6. Available: [Online] http://www.clove.co.uk/viewtechnicalinformation.aspx?content=3B2BD4 91–6465–4C70–ABDB–5A12A06C3D8D
7. Available: [Online]: http://en.wikipedia.org/wiki/CDMA2000
8. H. Hwang et al., "A Design of 700 MHz frequency Band LTE Receiver Front-end with 65 nm CMOS Process," Microwave Conference, 2009. APMC 2009. Asia Pacific, 7–10 Dec. 2009, pp. 720–723.
9. W. Y. Ali Ahmad, "Effective IM2 estimation for two-tone and WCDMA modulated blockers in zero-IF receivers", *RF Design*, pp. 32–40, April 2004.

Chapter 9
Tutorials

Abstract In this chapter, the student is taken through several tutorials, which provide him/her with an opportunity to synthesize the knowledge covered heretofore in the book. The tutorials presented are conducted with the Keysight SystemVue electronic system-level (ESL) design software. If a SystemVue license is not owned, the student may contact Keysight (please follow this link: http://www.keysight.com/find/eesof-radio-systems-engineering) for a time-limited evaluation license. In the tutorials, the student will gather practical experience with slightly simplified real-world design examples of radio transmission systems, both in the area of communications as well as radar sensing. The tutorials cover stateof- the-art system applications and consider the characteristics of typical radio-frequency hardware components. The tutorials will help you in:

- Getting a deeper understanding of the discussed theory by actively dealing with practical examples
- Getting a better impression of the performance of real radiofrequency hard-ware components and the persisting limits in system design
- Getting experience with a state-of-the-art software tool for communications system-level design

The tutorials can be worked through in parallel to the theoretical chapters of the book. For each tutorial the required prerequisites are indicated. The presented screenshots and results have been taken with the 2013.08 version of SystemVue. Future versions might have a slightly different layout and functionality.

9.1 Tutorial 1: Introduction to SystemVue and Basic Receiver Design

9.1.1 Learning Objective

The objective of this tutorial is to enable the reader to independently build a simulation model in SystemVue, parameterize the components, run the simulation, and read the important quantities from the simulation result.

H. J. De Los Santos et al., *Radio Systems Engineering,*
DOI 10.1007/978-3-319-07326-2_9, © The Authors 2015

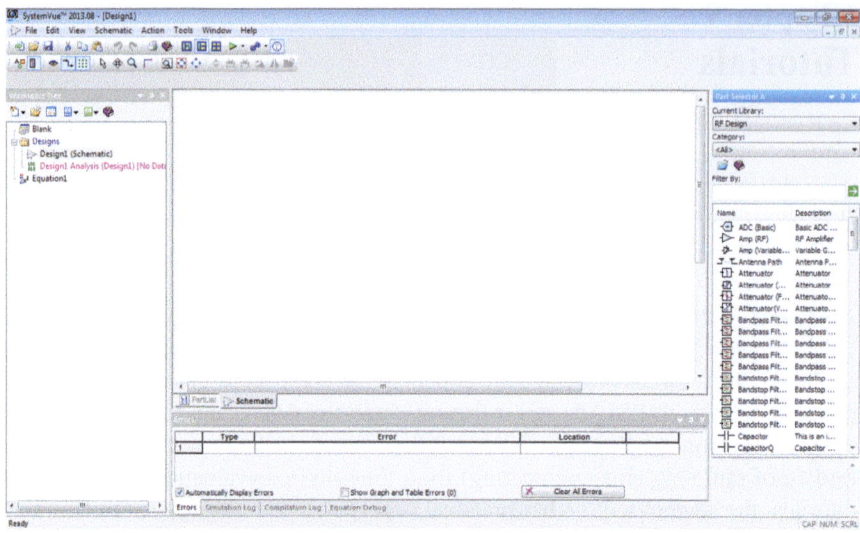

Fig. 9.1 New empty workspace in SystemVue

9.1.2   Prerequisites for this Tutorial

This tutorial requires a basic knowledge of properties of amplifiers and mixers, noise figure and signal-to-noise ratio, Friis' formula (*Chaps. 1–5*).

9.1.3   Design Task

A basic receiver for the down-conversion of an LTE radio signal to the base-band is implemented, considering the characteristics of real radio frequency hardware.

9.1.4   Getting Started and Schematic Window

After the startup of SystemVue, a window with a welcome message appears. Click on the *Close* button to get to another dialog window (*Getting Started with System-Vue—Please Select an Action*). In this window go to the section *Create a NEW workspace from a template* and select *Blank* to create a new empty design, then press *OK*. Maximize the *Design1* window in the center. Your screen should now look similar to Fig. 9.1.

In the *Workspace Tree* column on the left side, new windows can be created, which will become important later to display the simulation results. In the *Part Selector A* column on the right side the available component models are displayed. Under *Current Library* select *RF Design* to see the available RF component models. To get an overview of all available models you may browse through the complete list.

9.1.5   Implementation of Basic Receiver Model

To get started with SystemVue, we implement a basic receiver model consisting of an RF amplifier (LNA), a mixer for down-conversion of the received signal, a local oscillator to drive the mixer, and an IF amplifier to amplify the down-converted signal. The received signal from the antenna is simulated by a signal source component, which provides the signal with additive white Gaussian noise. To setup the design in SystemVue, the required components simply need to be dragged from the *RF Design* Library into the *Schematic* window and connected. The models used for the basic receiver model are the following.

Source (Multi) This component allows the generation of various signals (including also modulated signals) with additive white Gaussian noise and is typically used to simulate the signal received at the antenna.

Amp (RF) This model simulates the behavior of a real amplifier. All relevant parameters (e.g. gain, noise figure, 1 dB compression point, etc.) can be set by the user. This model can be used to simulate both RF and IF amplifiers.

Mixer With this model the behavior of a real mixer is simulated. Also, for the mixer, all relevant parameters (conversion gain, LO drive level, noise figure, etc.) can be provided/entered by the user.

Oscillator (Power) This component represents an oscillator that can be tuned to an arbitrary frequency. One typical application for this component is the simulation of a local oscillator to drive a mixer.

Output Port This component is required to terminate the RF model chain if no other terminating component is present.

These component models are used to set up the basic receiver model composed of RF amplifier, mixer, local oscillator, and IF amplifier. In this first step, do not make any modifications to the default settings. Your final design in the *Schematic* window should look similar to Fig. 9.2. If you take a closer look at the lines connecting the different models, you can see small green numbers, the so-called *net numbers*, which allow to distinctly access each part of the network when displaying the simulation results. Moreover, SystemVue automatically assigns numbers to the different components. All components considered as ports are assigned ascending numbers (1: MultiSource, 2: PwrOscillator, 3: Output) corresponding to the net numbers. Also, multiple instances of the same model are enumerated in ascending order (e.g. RFAmp_1, RFAmp_2).

9.1.6   Setting of Component Parameters

The component models contain default parameters, which now need to be adjusted to the given conditions. The individual settings of each model can be accessed by double-clicking on its icon. The *MultiSource* model has a more complex parameter menu, since it allows the simultaneous generation of multiple signals. To configure

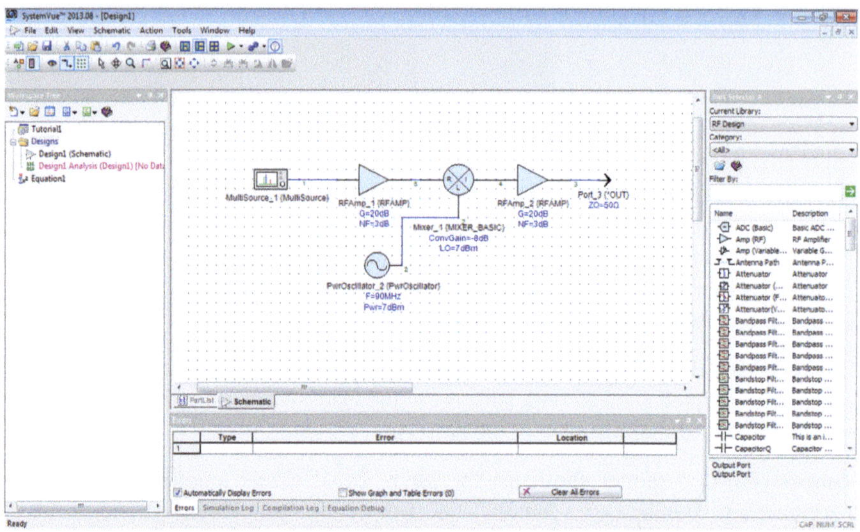

Fig. 9.2 Basic receiver model in schematic window with default settings

the *MultiSource*, double-click on the icon. Then create a new source by clicking the *Add* button in the *Source Summary* section. A new dialog window (*Edit Source*) will appear. In this window, under *Type*, select *Wideband*, enter a *Bandwidth* of 10 MHz and adjust *Number of Points* to 11. Under *Parameters,* set *Center Frequency* to 751 MHz and *Power* to −70 dBm. This roughly corresponds to a typical LTE signal. To simplify matters, the *Multi-Carrier* option is not used. If you confirm the settings by clicking *OK* in both windows, SystemVue automatically adds a short summary of the settings to the *MultiSource* icon. For the remaining component models, there is only a single level of settings, which is also accessed by double-clicking. For these models enter the values provided in the following. The values represent typical data of real RF devices for the regarded frequency range. If you tick/select the checkbox in the *Show* column behind the modified values, these values are also displayed below the model icon in the *Schematic* window. Your screen should finally show a design similar to Fig 9.3.

RFAmp_1 G=14.5 dB, NF=0.7 dB

Mixer_1 ConvGain=−9 dB, LO=13 dBm, NF=9 dB

PwrOscillator_2 F=741 MHz, Pwr=13 dBm

RFAmp_2 G=24 dB, NF=2.5 dB, OP1dB=16 dBm, OPSAT=20 dBm

9.1.7   Add RF System Analysis and Execute Simulation

Before the simulation can be executed, an *RF system analysis* needs to be added to the design and the signal path needs to be defined. To proceed, right-click on the

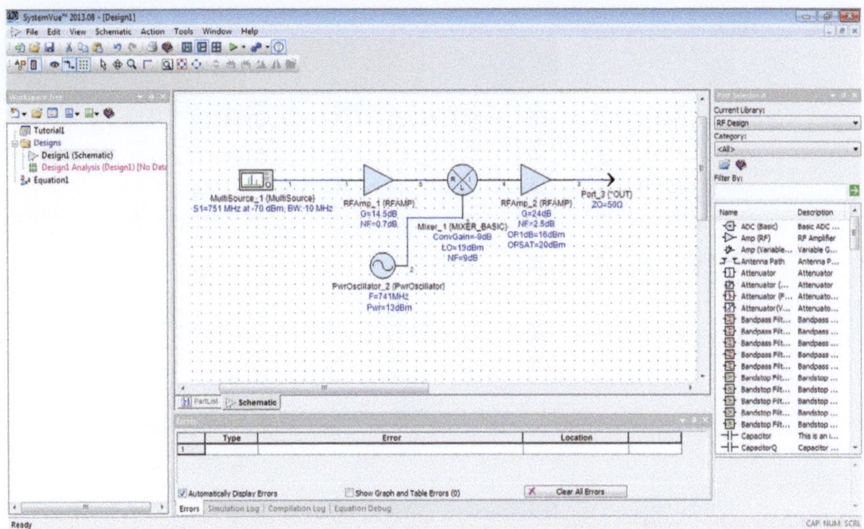

Fig. 9.3 Complete simulation model including component model parameters

Designs folder in the *Workspace Tree* column. Then, select *Add>> Analyses >> Add RF System Analysis* in the consecutively opening windows. A new window, entitled *System Simulation Parameters,* will open. In the section *Measurement Bandwidth,* adjust the *Channel* value to 10 MHz. In the section, *Schematic Source Summary,* both *MultiSource* and *PwrOscillator* should be automatically listed. To define the signal path, change to the *Paths* tab. By clicking the button *Add All Paths From All Sources,* SystemVue will automatically create a path from the *MultiSource* to the *Output Port.* Click on *OK* to confirm the settings and close the window. A new entry, *System1,* will appear in the *Designs* folder in the *Workspace Tree.* It is highly recommended to delete the *Design1 Analysis* entry from the *Workspace Tree* now, because, otherwise, the simulator might, additionally, start a *Data Flow* simulation, which would then produce an error message.

Congratulations! You are now ready to run your first simulation in SystemVue. You can start the simulation, either by clicking the *play* icon in the tool bar, selecting *Action >> Run All Out-of-Date Analyses and Sweeps* in the menu bar, or by simply depressing the *F5* key. When the simulation is finished, a new folder *System1_Data_Folder* and a new entry *System1_Design1_Data* will appear in the *Workspace Tree.* Also, in the *Errors* window, some additional information (hopefully no errors) might be displayed.

9.1.8   *Visualization of Simulation Results*

At the beginning, mainly two types of visualization are of interest. These are the *Spectrum Diagram* and the *Level Diagram.* With the *Spectrum* diagram, the signal spectrum can be observed at any point in the design. To create a new diagram (ei-

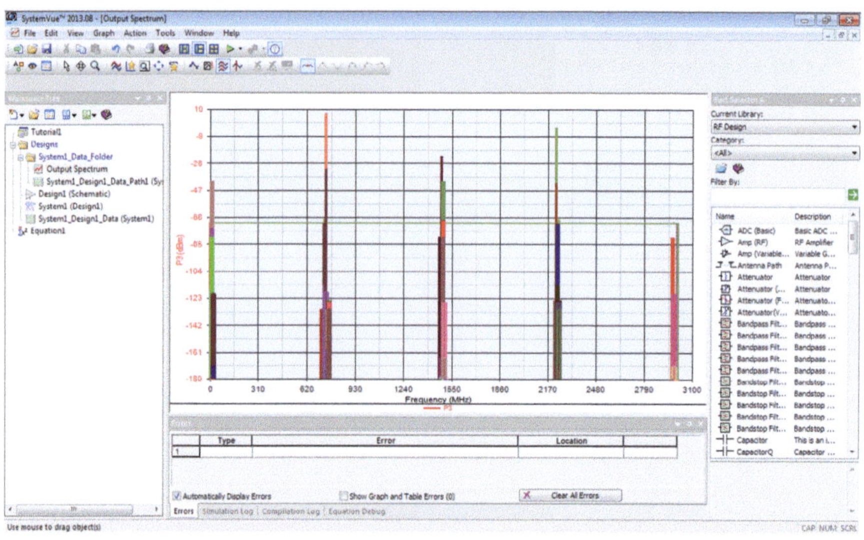

Fig. 9.4 Visualization of the output signal spectrum

ther *Spectrum* or *Level Diagram*) right-click on the *System1_Data_Folder* in the *Workspace Tree*. Then, select *Add >> Add Graph*. A new window with the title *Graph Series Wizard* will appear. To display the signal spectrum, select *Spectrum* in the *Select Type of Series* box. Then, in the *Select Data* box, all available spectra are displayed. You can choose, either power (P) or voltage (V) spectra for any net number of the design (number succeeding the letter). To display the power spectrum at the *Output Port*, which corresponds to net number 3, select *P3* in the *Select Data* box. Then, press *OK*. A new window that allows establishing the settings for the graph appears. To be able to distinguish the different graphs later, it is recommended to change the *Name* to, e.g., *Output Spectrum*. Optionally, you can make manual settings for the scaling of the axes in this window and add additional results to the graph. If you confirm the settings by pressing *OK*, your screen should look similar to Fig. 9.4.

As a first exercise, try to identify the original input signal, the LO signal, the down-converted signal, and the intermodulation products. You may use the *magnifier* tool from the toolbar to zoom in for a more detailed view. If you hold your mouse over spectral lines, SystemVue gives you additional information about their origin and power. With the default settings of SystemVue, only intermodulation products up to third order are simulated. As an additional exercise, you may also observe the spectrum at different points of the network, i.e., for different net numbers.

The creation of a *Level Diagram* is achieved in a very similar way. Again, right-click on the *System1_Data_Folder* in the *Workspace Tree* and select *Add>> Add Graph*. Now, under *Select Type of Series*, choose *Level Diagram*. In the *Select Diagram* box, various quantities will appear. Three very important quantities are discussed in the following. For additional information, please refer to the SystemVue help.

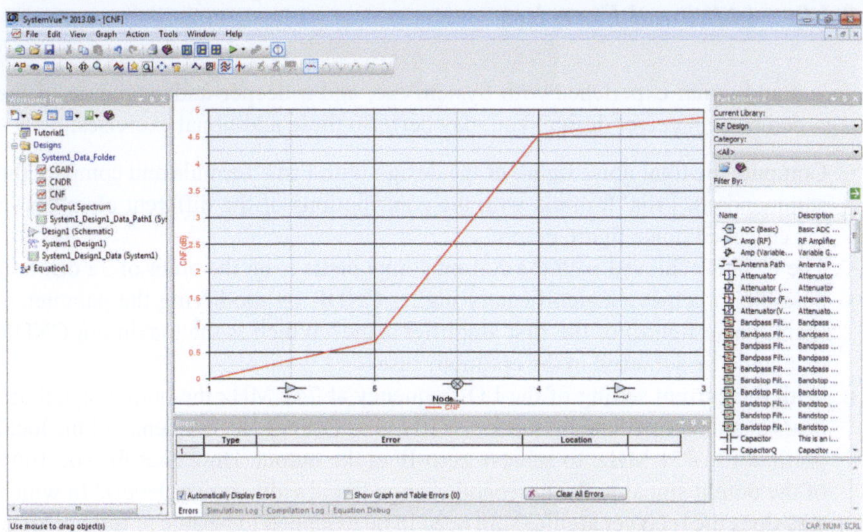

Fig. 9.5 Level diagram of the cascaded noise figure (CNF)

CGAIN (Cascaded Gain) Displays the cascaded gain from the input to the output.

CNDR (Carrier to Noise and Distortion Ratio) Describes the ratio of the desired signal and any unwanted components, and practically represents the signal-to-noise ratio.

CNF (Cascaded Noise Figure) Displays the cascaded noise figure along the signal path.

Create one individual graph for each of these quantities. Do not forget to adjust the name of each graph in order to be able to distinguish them easily in the *Workspace Tree*. Figure 9.5 shows a screenshot of the CNF plot, as an example. In each of the level plots you can follow the investigated quantity through the entire network, from the signal source to the output. This is one of the most powerful tools of SystemVue for the analysis of a system design. In each graph, the abscissa shows the net numbers along the defined path, including icons of the different components. The ordinate shows the total value of the regarded quantity accumulated over the different stages. In the example shown in Fig. 9.5, you can see how the cascaded noise figure increases according to Friis' formula. The noise figure of the first stage, 0.7 dB, fully contributes to the cascaded noise figure; the contribution of the following stages is clearly below their actual noise figure. With SystemVue, the impact of different component parameters on the total system behavior can be investigated in a very convenient way. That is, you can simply modify the gain or the noise figure of the first stage, re-execute the simulation, and all graphs will automatically update and show the new results. You may try this as an exercise.

9.1.9 *Additional Exercises*

To get additional experience with SystemVue, and a deeper understanding of the regarded basic receiver design, you may perform these additional exercises:

- Compute the total noise figure of the design with Friis' formula and compare the summands in Friis' formula with the contributions of the different elements in the cascaded noise figure graph.
- The CNDR achieved with the current components is on the order of 32 dB. Is it possible to achieve a significantly higher CNDR by modifying the parameters (gain, noise figure) of the first amplifier stage? Which is the maximum CNDR that could be achieved in the optimum case?
- With the current setting of the LO frequency at 741 MHz the output signal appears at an intermediate frequency of 10 MHz. Change the frequency of the local oscillator to 751 MHz, to achieve zero-IF at the output. How does the spectrum of the output signal in the baseband change (bandwidth, power level)? In which way does the receiver architecture need to be extended to be able to fully recover the information in the zero-IF signal?

9.2 Tutorial 2: Direct Conversion Receiver Design

9.2.1 *Learning Objective*

In this tutorial, the reader will strengthen his/her knowledge of the working principle of a direct conversion receiver and learn how to choose appropriate components.

9.2.2 *Prerequisites*

This tutorial requires knowledge of I/Q-demodulation, bandpass and lowpass filters, power splitters (**Chaps. 1–7**).

9.2.3 *Design Task*

A direct down-conversion receiver for the LTE standard, including an I/Q-mixer and filters is designed.

9.2.4   Implementation of Direct Down Conversion

The simulation model of the basic receiver from Tutorial 1 is now extended towards a direct down-conversion architecture. A screenshot of the first implementation step is shown in Fig. 9.6. However, it is recommended that you try to implement the architecture on your own. You can start with the file from Tutorial 1 and modify it step by step. In the following, some important advice will be given.

Required Components All components that are part of the basic receiver design (source, amplifiers, mixer) can be reused to build the direct down-conversion receiver. If multiple instances of a certain component are required, you can simply clone a component, including its settings, by a simple copy and paste operation. To optimize the layout, and avoid crossing signal paths, it is possible to rotate and mirror all components (via the toolbox).

Power Splitters The only new component type that is required for the direct down-conversion architecture, are power splitters. They are both necessary for splitting the RF input signal and splitting the local oscillator signal. There are also power splitters with included phase shifters available, which can be used for a convenient implementation of the splitting of the LO signal.

Local Oscillator Settings Since the aim is now to bring the signal directly into the baseband (zero IF), the frequency of the local oscillator must be changed to the frequency of the source (751 MHz). Also, please keep in mind that the power of the LO signal is reduced by the splitter and, therefore, not sufficient to drive the mixer anymore. Adjust the LO power settings accordingly (3 dB increase).

Multisource Settings For compliance with the LTE standard, the receiver needs to be able to operate with a minimum received signal power of -90 dBm. Therefore, the signal power of the multisource is now adjusted to -90 dBm.

Simulation Settings The design has now two output ports (I and Q channels). Therefore, the settings for the signal paths need to be adjusted. To add the additional path, double click on the *System1* icon in the *Workspace Tree* and select the *Paths* tab. The most convenient way is to delete the existing path and then use the *Add All Paths From All Sources* wizard.

The model of the direct down-conversion receiver is now ready for a first simulation. To be sure that your own design is correct, you may now compare it with Fig. 9.6.

9.2.5   First Simulation and Analysis of Results

Now, please execute the simulation. The graphs, that have already been defined in the *Workspace Tree*, will automatically update. Moreover, *SystemVue* will issue the information that the lowest channel frequency is below 0 Hz. For that reason, also

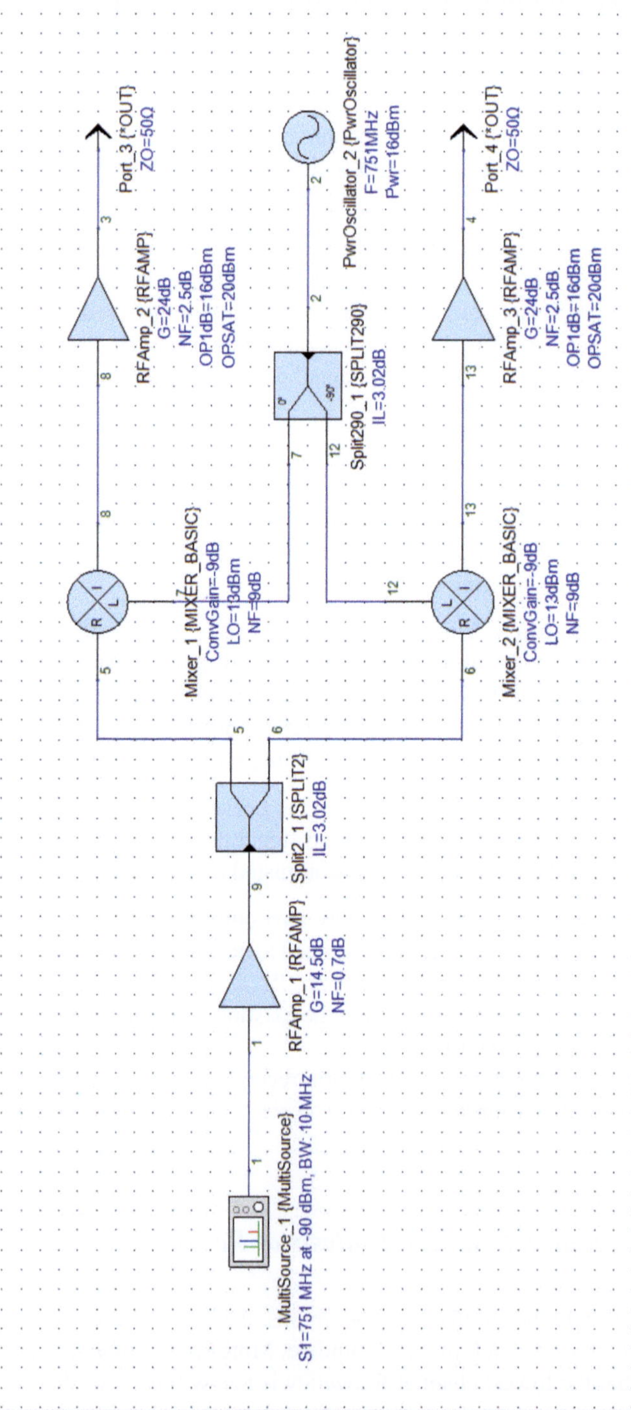

Fig. 9.6 First implementation step of direct down-conversion architecture

the computation of the channel power fails and, as a consequence, the CNDR plot shows incorrect results. It is, therefore, recommended to delete the CNDR plot from the *Workspace Tree*. Performing the following analyses will help you understanding the properties of the direct down-conversion concept and give you an idea of why additional components are required to get a fully working design.

- Select the output spectrum graph from the *Workspace Tree* (if you have not carried over the graphs from Tutorial 1, simply generate a new one). You may also add a second graph for the additional output port and convince yourself that the power spectra of both ports are identical. Take a look at the different components of the output spectrum, and try to find explanations as to where and how they are generated. How can you improve the design so it can suppress the undesired harmonics of the source and the LO signal?
- The most interesting area of the output spectrum is the desired signal which, theoretically, spans from -5 to $+5$ MHz. However, as already mentioned previously, the RF system analysis of *SystemVue* is not able to consider negative frequencies. To adjust the scaling, simply click on the x-axis and uncheck *Auto-Scale* in the *Properties* window. Then, set the displayed range from 0 to $+10$ MHz.
- Try to identify the down converted signal from the source in the output spectrum. It should range from 0 to $+5$ MHz, the negative and positive frequencies overlap.
- The SNR of the received signal can be determined by comparing the source signal power with the node noise power, both measured at the IF amplifier. With the given settings, you should observe an SNR of approximately 9 dB.
- One drawback of the direct down-conversion architecture is the DC offset in the output signal resulting from the self-mixing of the local oscillator. Since the desired signal converted to the baseband contains positive and negative frequency components, the DC offset appears in the center of the bandwidth of the desired signal. Compare the power level of the DC component (0 Hz) with the power of the desired signal and determine the ratio. Is there any possibility of completely suppressing the DC offset or can this only be achieved by changing to another receiver architecture? Is it possible to, at least, lower the relative level of the DC offset?

9.2.6   *Optimized Implementation*

The analysis of the simulation results has shown that certain modifications are required. This mainly concerns the following two issues.

- Filters need to be integrated to suppress the undesired harmonics at the output.
- An additional amplifier stage needs to be integrated to increase the power level of the desired signal.

For the new components that are now added to the design, the following settings are chosen.

RF Filter A first filter is inserted directly after the receiving antenna to avoid undesired intermodulation products from sources with other carrier frequencies. For that purpose, the component *Bandpass Filter (Chebyshev)* is chosen. Set the *lower edge frequency* to 670 MHz and the *higher edge frequency* to 795 MHz (this is a typical example of a commercially available device, there needs to be some room for tuning the receiver to different carrier frequencies). For the *implementation loss* a value of 2.2 dB is assumed and the *filter order N* is set to 3.

IF Filter The task of the IF filters is to suppress all frequencies in the down-converted signal that are higher than the source bandwidth. The IF filters are placed in each branch after the mixer. The corresponding component is *Lowpass Filter (Chebyshev)*. Since the bandwidth of the signal does not change, the *passband edge frequency* is set exactly to the maximum frequency of the IF signal (5 MHz). For the IF filter, an *implementation loss* of 1 dB is assumed and the *filter order N* is again set to 3.

Second RF Amplifier To increase the power level of the desired signal, a second RF amplifier stage is inserted directly after the first one. This helps reducing the impact of the DC offset and might also be required to provide sufficient signal power for the A/D converters. Since the cumulative noise figure is mainly determined by the first amplifier stage, for the second stage typically an amplifier with higher gain and higher noise figure is chosen. For the simulation, a gain of 23 dB and a noise figure of 1.7 dB are assumed.

The complete simulation model is shown in Fig. 9.7. Execute the simulation and open the output spectrum graph. First have a look at the entire frequency range. Are the intermodulation products suppressed? Now change again to the detailed view from 0 to 10 MHz. Can you see an impact of the additional components on the SNR at the output? How have the output signal power and the relative level of the DC offset changed?

9.3   Tutorial 3: Heterodyne Receiver Design

9.3.1   Learning Objective

In this tutorial, the reader will understand the differences between a direct-conversion receiver and a heterodyne receiver regarding implementation and performance, and will be able to choose the appropriate concept for a given application.

9.3.2   Prerequisites

This tutorial requires knowledge of I/Q-demodulation, heterodyne receiver architecture, bandpass and lowpass filters, power splitters *(Chaps. 1–7)*.

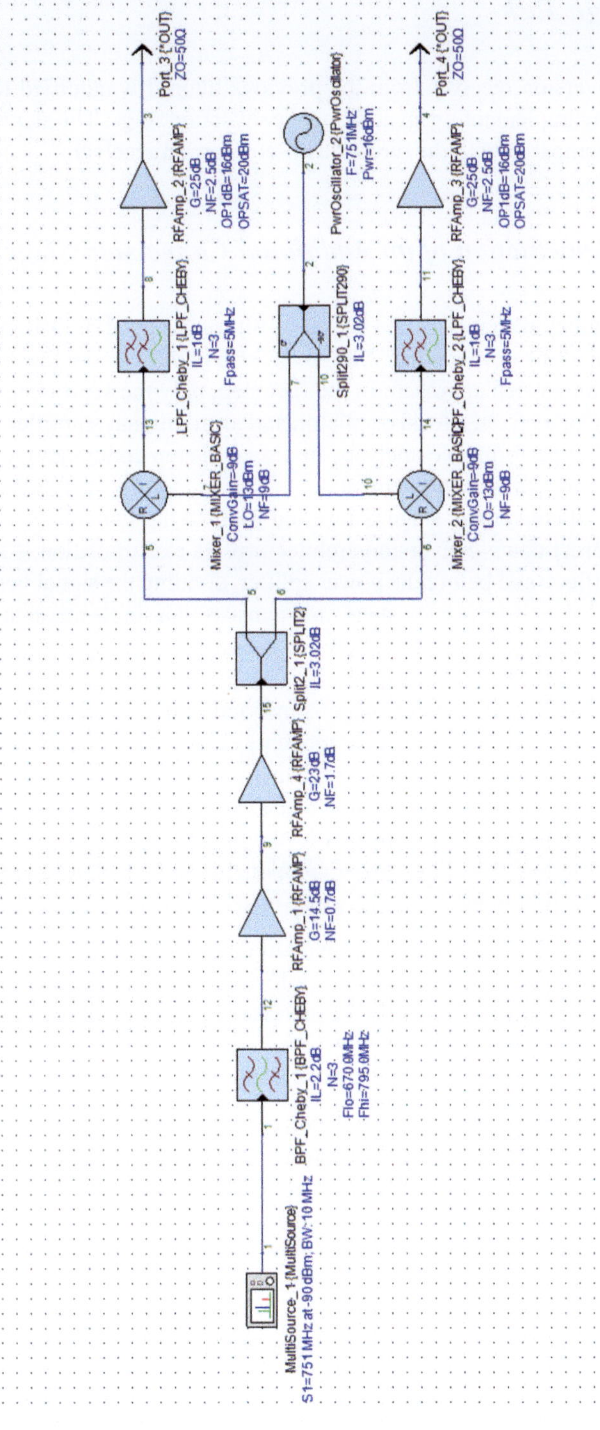

Fig. 9.7 Complete implementation of direct down-conversion receiver

9.3.3 Design Task

The direct-conversion receiver from the previous tutorial is modified to a hetero-dyne receiver by implementing an additional conversion stage. The signal quality is analyzed and compared with the direct down-conversion receiver.

9.3.4 Implementation of an Additional Conversion Stage

Again, the simulation model of the previous tutorial is used as the basis for the implementation of the new concept. Moreover, also in this tutorial the implementation is done in a two-step approach. In the first step, only the additional conversion stage consisting of a mixer, a local oscillator, and an additional amplifier is inserted. Then, after a first performance analysis, in a second step additional modifications are made for performance improvement.

The additional conversion stage is inserted prior to the I/Q conversion stage of the available direct down-conversion architecture. Its exact position is subsequent to the RF amplifiers, and before the power splitter, dividing the signal in the I and Q branch. The additional conversion stage only consists of a single mixer and does not require splitting into an I and Q branch, since the output signal is at an intermediate frequency and not in the baseband.

When the additional conversion stage is inserted, it must be considered that the signal frequencies in the subsequent components significantly change, with respect to the original direct down-conversion implementation. Therefore, even if the component architecture of the I/Q conversion stage can be adopted without changes, it is necessary to adjust certain parameters of the components to the new operation frequency. All required modifications of components and parameters are explained in the detail in the following.

Additional Mixer The new mixer of the additional conversion stage converts the signal from the carrier frequency to a very low intermediate frequency, which is very close to the baseband. Therefore, the mixer model can be copied from the direct down-conversion receiver without any parameter changes (ConvGain$=-9$ dB, LO$=13$ dBm, NF$=-9$ dB). The mixer is inserted after the second RF amplifier.

Additional Local Oscillator To drive the mixer of the additional conversion stage, a second local oscillator is required. You may copy and paste the LO from the I/Q stage. However, now the settings have to be adjusted. Since the LO signal is not split in this case, the LO power has to be set to the input power required by the mixer (13 dBm). Moreover, the LO frequency needs to be adjusted to convert the signal to an intermediate frequency. For this exercise, an intermediate frequency of 10 MHz is chosen, similar to the basic receiver in Tutorial 1. Therefore, the LO frequency of the first conversion stage needs to be set to 741 MHz.

Additional Amplifier As a result of the high insertion loss of the additional mixer, and the very low input signal power, the simulation will not converge, unless an additional amplifier stage is inserted subsequent to the new mixer. This amplifier is operating at the intermediate frequency. Its parameters are set to G=20 dB and NF=1.7 dB.

Mixers of I/Q Conversion Stage The input frequency of the mixers of the I/Q conversion stage has now changed by decades. Therefore, the parameters of the mixers must be adjusted to reflect the realistic behavior of components designed for that frequency range. The new parameters for the mixers of the second stage are: ConvGain=−5 dB, LO=7 dBm, NF=9 dB.

Local Oscillator of I/Q Conversion Stage Also, the settings of the local oscillator of the second stage need to be adjusted, as a result of both the changed input frequency and the modified input power of the mixers. The signal now needs to be down-converted from the intermediate frequency of 10 MHz to the baseband. Therefore, the LO frequency is set to 10 MHz. The output power is reduced to 10 dBm to provide an input power of 7 dBm (after splitting) to each mixer.

Now, all necessary modifications have been made to get an operational implementation of a heterodyne receiver in SystemVue. To make sure that your design is correct, you may now compare it with the screenshot shown in Fig. 9.8. Run the simulation and perform the following analyses:

- Observe the progress of cascaded gain and cascaded noise figure over the different components.
- Analyze the output spectrum both in the overall view as well as in a detailed view from 0 to 20 MHz.
- Determine the SNR of the output signal and compare it with the result obtained for the direct down-conversion receiver.
- The SNR should have significantly reduced compared to the direct down-conversion receiver (by approx. 5 dB). Try to find the reason for that reduction.
- For that purpose, it is helpful to observe the spectrum also at other nodes within the network. Create a new spectrum plot for the output signal of the first conversion stage (e.g., at the output of the first mixer) and analyze the different signal components.
- Which additional system component is required at this point to improve the performance of the receiver?

9.3.5   *Optimization of the Design*

The analysis of the IF and the output spectrum has shown that the signal contains undesired spectral components. To suppress these components, an additional bandpass filter must be inserted in the IF stage, between the IF mixer and the IF amplifier. For the IF filter, the following parameters are chosen.

IF Filter Filter type *Bandpass Filter (Chebyshev)*, with the following parameters IL=1.5 dB, N=3, Flo=5 MHz, Fhi=15 MHz

The final optimized design is shown in Fig. 9.9. Run the simulation for the optimized design and perform the following tasks:

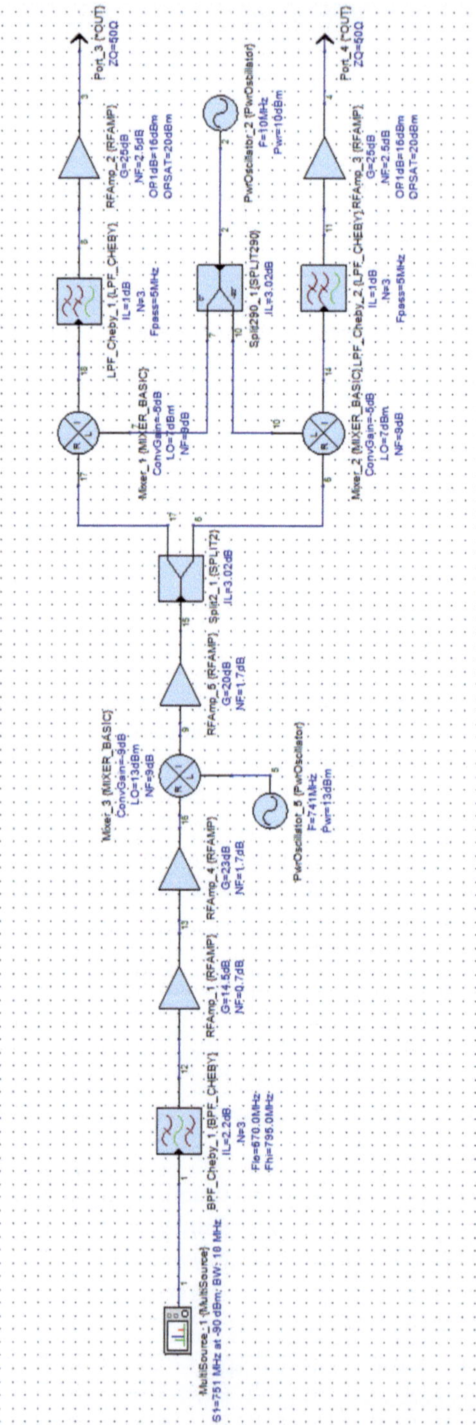

Fig. 9.8 First implementation step of the heterodyne receiver

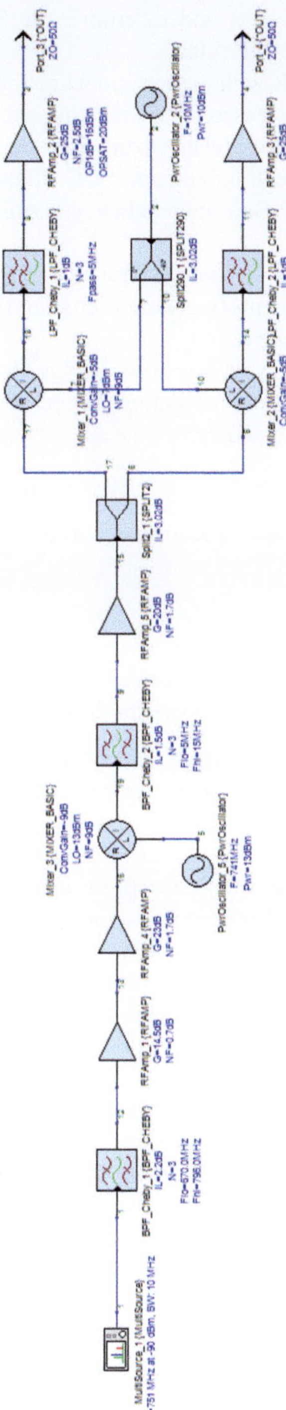

Fig. 9.9 Final optimized design of the heterodyne receiver

- Analyze the simulation results and determine the SNR at the output that is achieved with the additional IF filter.
- Compare the achieved SNR with the result obtained for the direct conversion receiver. Can you explain the reason for the remaining difference? Is this a true performance degradation or, rather, related to the signal representation in SystemVue?
- Does the output signal spectrum contain a DC offset like in case of the direct down conversion receiver? Does it contain any intermodulation products of significant power level?
- Do you see possibilities for further performance improvement? Modify the parameters of certain components (e.g. amplifier gain, filter edge frequencies) and observe the impact on the SNR.
- As an additional exercise, you may also change the carrier frequency of the source. Which of the two local oscillators do you need to adjust to tune the receiver to the new carrier frequency?

9.3.6   *Comparison of Direct Down-Conversion and Heterodyne Receiver*

Both the direct down-conversion architecture and the heterodyne receiver principle have their advantages and drawbacks. Answering the following questions will give you an understanding of the proper choice for different applications:

- Which is the difference in complexity and the amount of required system components? Which concept would you prefer if you needed to build a cost-effective receiver?
- How do the different receiver types differ regarding the spectral purity of the output signal? Which is the better concept for receiving signals with a modulation of high order (e.g., QAM)?
- If you only transmit at low data rates with a robust modulation technique, is there any benefit from spending additional effort for a heterodyne receiver or would you expect comparable performance with a direct down-conversion receiver?
- Does one of the concepts reduce the effort of the subsequent digital signal processing for demodulation (think of the DC offset)?
- If you need to be able to tune to different carrier frequencies over a wide range, does one of the concepts offer advantages?
- There might be other strong signals on neighboring channels, which you do not want to interfere with the desired signal in the receiver. Which concept would give higher robustness in this case?

9.4 Tutorial 4: PSK Transmitter Design and Complete RF Transmission Link

9.4.1 Learning Objective

In this tutorial, the reader will deepen his/her knowledge of digital modulation and will get an impression of the receiver architecture requirements depending on the modulation technique used.

9.4.2 Prerequisites

This tutorial requires knowledge of modulation and frequency conversion schemes. In particular, knowledge of QPSK, 8-PSK and 16-QAM is necessary *(Chaps. 1–7)*. Furthermore, topics such as inter-symbol interference, aliasing and bit error rate will be indirectly addressed and will be assumed known by the reader.

9.4.3 Design Task

In this tutorial, a PSK transmitter for the LTE standard is implemented in a direct up-conversion architecture. The transmitter is combined with the receiver from Tutorial 2, and the performance of the complete link is investigated.

9.4.4 SystemVue Data Flow environment

Up to this point, only elements of the SystemVue RF Design library have been used. Because of this, simulations have been done so far with the *Spectrasys* RF system simulator embedded into SystemVue. This analysis engine utilizes a spectral-domain simulation technique called SPARCA—Spectral Propagation And Root Cause Analysis. This means that all previous designs were analyzed within the frequency domain. In order to study the performance of communication systems at the algorithmic level, however, time-domain simulations are needed. Therefore, SystemVue also provides users with a time-domain simulation engine called Data Flow. This is actually the main simulation engine within SystemVue, and is used, by default, with almost all the component libraries. The only exception is the *RF Design* library, for which the Spectrasys engine is automatically activated. In the following sections, unless otherwise stated, the Data Flow environment will be used.

9.4.5   Implementation of Direct Up Conversion

A simulation model of a direct up-conversion architecture will be set up. This will be the complementary model to that of Tutorial 2, i.e., a 751 MHz carrier frequency and 10 MHz signal bandwidth will be considered. A screenshot of the simulation model is shown in Fig. 9.10. The setup of the simulation model is done in the same manner as with the components of the *RF Design* library. In the *Part Selector A* column, on the right-hand side under *Current Library*, select *Algorithm Design* to see the most important Data Flow components. In the following, the configuration of all necessary components for an LTE transmitter will be discussed.

Random Bit Generator This component generates a random bit sequence of 1 s and 0 s to be used for testing a communication system. In this case, the default settings will be left untouched.

Gray Encoder As its name implies, this component converts a certain bit sequence into a gray-coded bit sequence. The key parameter to be set is the number of bits, i.e., *NumBits,* which will be set to 2 for the QPSK modulation scheme to be implemented here.

Complex Symbol Mapper This component maps a sequence of bits into a complex signal based on the modulation scheme chosen. Different modulation schemes map the input bits differently in the complex space, thus forming different constellations. For the case to be considered in this part of the tutorial, the parameter *ModType* will be set to QPSK.

CxToRect This model converts input complex values to output real and imaginary values. In this way, the information signal contained in both I and Q components can be individually filtered.

Raised Cosine Lowpass Filter (Generic Filter Design Model) Lowpass filter model in order to improve the spectral efficiency of the signal, i.e., to eliminate high frequency components. As it has been explained previously, failing to filter the transmitted signal would require a non-feasible infinite bandwidth. The component is generated by accessing the generic filter generator and choosing *Lowpass* as *Response* type and *Raised Cosine* as *FIR* type. In addition, *SymbolRate* should be set to 3 MHz, *RollOff* to 0.35 and *LengthOption* to *1:Number of Taps*. With the *LengthOption* now modified, the parameter *Length* becomes available and should be set to 129. Finally, the parameter *Interpolation* is set to 20. Here, *SymbolRate* has been chosen in such a way that the signal bandwidth fits into the available channel bandwidth.

Modulator With this component, the signal is up-converted in frequency for transmission at the desired carrier frequency of 751 MHz. This supposes changing *FCarrier* to 751 MHz, whereas *InputType* should be left to its default value of *0:I/Q*.

Sink At each point where the signal will be observed, a sink needs to be added. This is the Data Flow equivalent of the output ports used in other tutorials. The

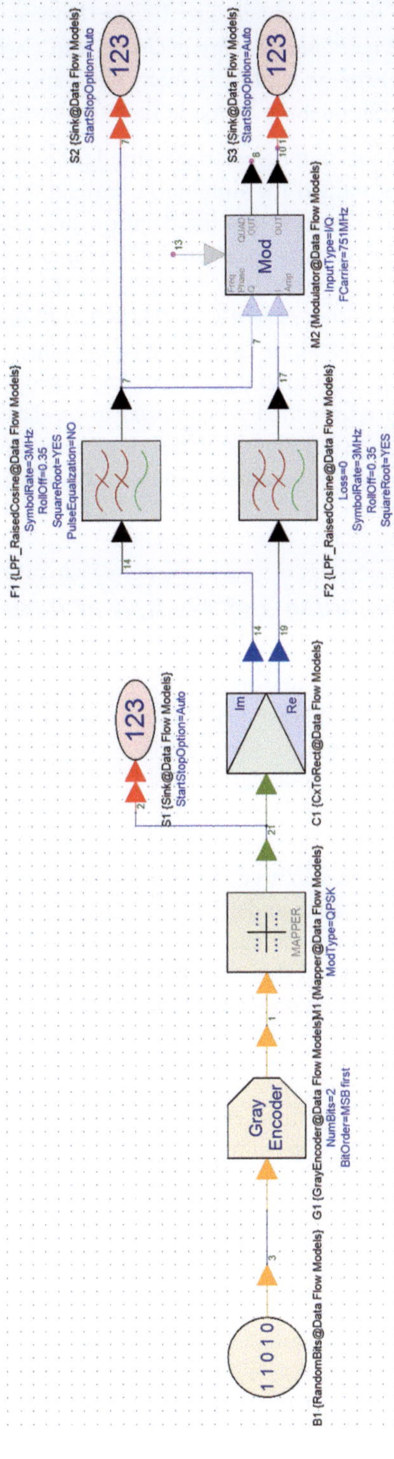

Fig. 9.10 Transmitter model in SystemVue

parameterization options of this component are diverse. However, the most important ones are the options *Create and Display a Graph* and *Create and Display a Table* found under the tab *Graph and Table*. In many simulations, it makes sense to generate a graph or a table. For the time being, both options will be left unchecked.

Simulation Settings In contrast to the Spectrasys simulation engine, for Data Flow models no paths need to be defined for simulation, instead the simulation timing settings have to be specified. For this, first the dataflow analysis has to be added. To proceed, right-click on the *Designs* folder in the *Workspace Tree* column. Then, select *Add>> Analyses >> Add Data Flow Analysis* in the consecutively opening menus. A new window, entitled *Data Flow Analysis*, will open. In the section, *Default Source and Sink Parameters for Data Collection*, adjust the *System Sample Rate* to 6 MHz and the *Stop Time* to 200 µs. Here, the choice of Sample Rate is based on the bit generation rate, which is supposed to be the smallest time unit within the model, whereas the stop time is chosen long enough to have sufficient samples for analysis. The rest of the parameters in this section will update automatically and should be left untouched.

9.4.6 Transmitter Simulation and Analysis of Results

Once the model is finished, the first step, as with the previous tutorials, is to execute the simulation. This way, the variables and signals are updated. Once the simulation is finished, new graphs will be created. First, the frequency spectrum at the output of the filter and at the output of the modulator will be plotted. Then, the timed signal at the output of the filter will be included. Finally, an eye diagram for this communication system will be added. In the case of the eye diagram, the timing criteria for the measurement needs to be specified, i.e., the symbol rate and the number of cycles before wrapping. This is done under *Custom Equations* in the *Graph Series Wizard* windows that appears when adding a new graph to the workspace or when editing a graph series. In the case of this tutorial, SymbolRate must be set to 3E6, i.e., 3 MHz, and NumCycles to 2. With help from the resulting graphs, perform the following analysis:

- Select the output spectrum graph from the *Workspace Tree* and compare it with the spectrum at the output of the filter. First, notice that there are no longer individual components, as was the case with the Spectrasys simulation. How large is the signal bandwidth and how is it related to the bit rate and the symbol rate?
- In addition to the graphs mentioned above, there are other important graphs that can be generated using SystemVue. One such graph is the constellation plot. Please, plot the constellation plot from this transmitter. Is the constellation plot in accordance with an ideal QPSK constellation? Can you also visualize the constellation diagram of the output signal? Why are the symbols in the output signal distorted with respect to the constellation diagram at the output of the mapper? Hint: Play with the filter parameters, in particular, the roll-off factor, to see how

this affects the constellation plot. Remember that, not having a filter at all would require a transmission channel of infinite bandwidth, which is not feasible.

- Previously, you were instructed to generate an eye diagram for this transmitter. However, no mention was made as to which signal should be used in order to generate the eye diagram. With which signal did you generate the eye diagram and why? Is there any difference between the eye diagram of the I and Q channels? Why?
- Change the modulation type to 8-PSK and 16-QAM, and observe the changes. How does the eye diagram change?

9.4.7   Receiver Integration

To evaluate the performance of the receiver, the already existing direct down-conversion receiver RF model from Tutorial 2 is integrated into the simulation. This supposes mixing up two different simulation setups, the Spectrasys one, from Tutorial 2, and the Data Flow one, used for the transmitter above. This approach allows the design of communication systems at the algorithmic level, but taking into account RF effects such as reverse transmissions due to non-ideal mixing of the RF signals. In order to do so, open both workspaces (Tutorial 2 and 4) simultaneously, and copy the receiver schematic from one workspace into the other. Once the schematic has been copied into the current workspace, close the workspace of Tutorial 2 and open the transmitter schematic. Then, from the *Workspace Tree* column, drag the receiver schematic you just copied into the current view and an *RF_Link* component will be automatically generated. This *RF_Link* encompasses the receiver sub-network and should be added after the modulator. In this fashion, the receiver model remains linked to this component, so that any changes on the receiver will be automatically taken into account.

In order to test the complete transmission chain after the RF_Link component, the additional demodulation stages will be added and configured as follows. The resulting demodulation chain can be seen in Fig. 9.11.

Raised Cosine Lowpass Filter (Generic Filter Design Model) Once the signal has been down-converted, it is once more lowpass-filtered, with the exact same settings used for transmission.

RectToCx This model converts the single I and Q paths into a complex signal, necessary for the following demapper.

Complex Symbol Demapper This complementary component to the mapper, used for transmission, maps a complex signal into a sequence of bits based on the modulation scheme chosen. As done previously, the parameter *ModType* will be set to QPSK.

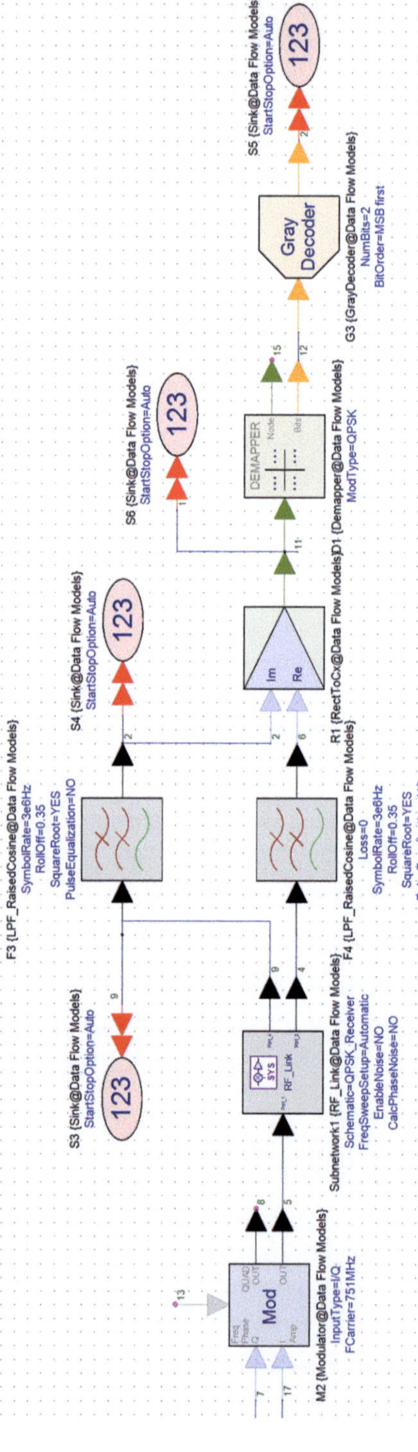

Fig. 9.11 Receiving chain in SystemVue

Gray Decoder   Analogously to the encoder above, this component converts a gray-coded bit sequence into a regular one. The key parameter to be set is the number of bits, i.e., *NumBits,* which will be set to 2 because of the QPSK modulation.

RF_Link, Receiver Schematic   As last step in the setup of the demodulator chain, all filter stages from the receiver will be shorted out in order to simplify the analysis. The resulting schematic should look like that in Fig. 9.12.

9.4.8   Signal Recovery Analysis

Once the initial transmit sequence has been recovered, a very quick way to verify the design is to take a look at the received constellation graph. Even though this is no assurance that the right bits were recovered in each case, based on the location and spread of the received signals the proper functioning of the receiver design can be inferred. Another option is to take a look at the eye diagram of the received signal. In the case of the QPSK modulation scheme, this means looking at the eye diagram of the single I and Q channels. If the eye shape is preserved, then the transmission and reception were successful. Most commonly, however, the eye diagram will have experienced some kind of distortion leading to its closing.

In spite of the very visual feedback that both the constellation graph and the eye diagram provide, an evaluation of the frequency spectrum is always of great assistance when analyzing communication systems. With this understanding, try to answer the following questions:

- Generate a timed signal graph of either the I or Q channel. Now compare it with the timed signal graph of the same component at the transmitter. What differences do you observe? Why does the signal exhibit an offset with respect to the original one? Is there any way to mitigate this effect? Verify that your proposed measure would in effect reduce the signal offset at the receiver. Hint: Take a look at the spectrum graph of the received signal and concentrate on the lower spectral components.
- Which non-ideal behaviors of the RF components can you observe, based on the constellation graph? How can these effects be mitigated?
- How large is the bandwidth of the down-converted signal in the baseband? Why? Based on this, what is the difference between plotting the eye diagram before and after the filter stage of the receiver? How do you expect the eye diagram to look in each case? Verify this with the simulation. Hint: There are two effects to be noticed.
- Once again, try out different modulation schemes and verify the proper functioning of each one of them. Based on the current receiver, which modulation schemes deliver the best results?

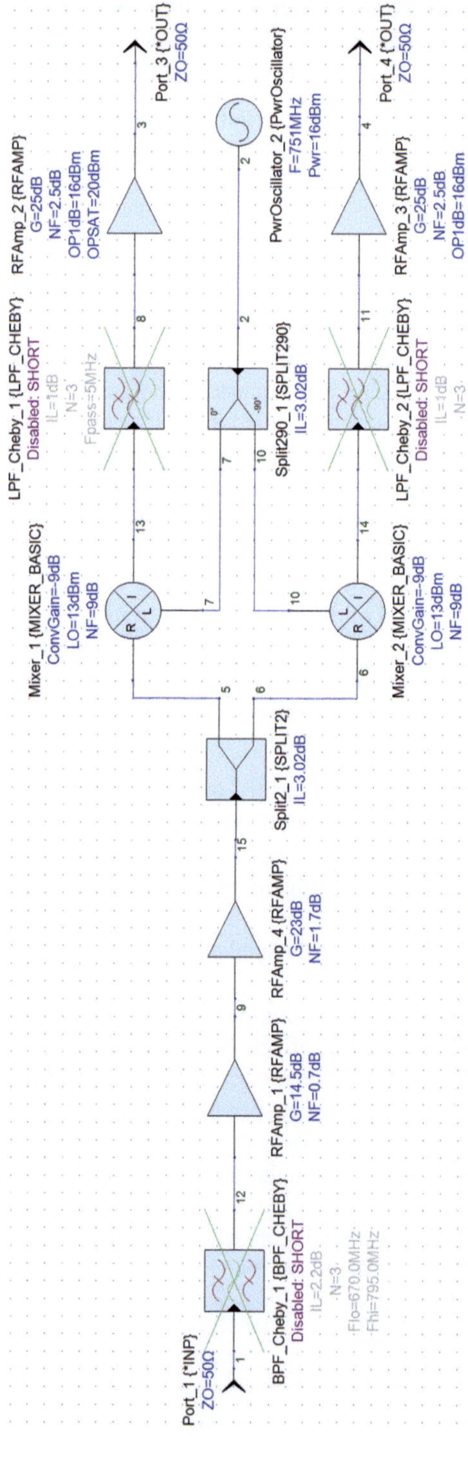

Fig. 9.12 Receiver model in SystemVue with filtering stages shorted out

9.4.9   Bit Error Rate Simulation

Up until now, evaluation of the modulation scheme has been greatly qualitative. In real communication systems, quantitative measures are also needed in order to benchmark the performance of different modulation schemes and algorithms; as was introduced previously in the book, the most important measure used in communication systems is BER (Bit Error Rate). In order to simulate the BER of our QPSK system, the receiver chain will be modified to include noise effects and reduce simulation time. Furthermore, some key parameters will be parameterized.

The first step will be to explain how a certain model can be parameterized in order to tune or sweep certain parameters. The easiest way to do this is by selecting the tune option for the desired parameter in the properties of each component. There are, however, parameters that propagate through components such as the sample rate. In this case, it is better to define general parameters through the equation sheets. For this, go to the *Designs* folder in the *Workspace Tree* column and right-click your mouse on it. Then, choose *Add >> Add Equation* in the consecutively opening menus. An equation takes the form of a script (just like the ones within the Matlab environment, for those familiar with Matlab), and just like in Matlab, defining a variable requires only an $a = b$ expression. In this way it is possible to have a globally defined variable that can be used from within different components while being defined only in one place. In order for it to be tunable, however, a slightly different expression has to be used, namely, $a = ?b$. Given the similarity to other scripting languages, no further details will be given here. The interested reader can search for more information regarding the scripting language of SystemVue within the help menu. For the BER simulation, the script shown in Fig. 9.13 will be used.

In this SystemVue equation, the most relevant variables have been included. In this manner, all numerical parameters can now be replaced with the variable names used in this equation, and the schematics will be updated automatically every time the equation is changed. Furthermore, it is seen that the *Rolloff*, *Symbolrate* and *EbNO* variables have been made tunable. In the following, some of these variables will be used to finalize the simulation setup of Fig. 9.14 necessary to determine the BER curve of this QPSK system.

Envelope to Complex Given that the joint simulation of Spectrasys and Data Flow components takes a larger toll in simulation time and memory the RF_Link will be replaced and down conversion will be done with this component which ideally converts any RF or IF signal to baseband. For this component no parameters are needed.

Raised Cosine Bandpass Filter (Generic Filter Design Model) In addition, since the signal is no longer divided into I and Q channels for down conversion the filtering is performed at the RF frequency and thus a bandpass filter is used. Configuration of the bandpass filter is analog to that of the lowpass filter used so far. This filter is generated by adding a generic filter component to the schematic and choosing *Bandpass* as *Response type* and *Raised Cosine* as *FIR* type. In addition,

```
1    Rolloff =? 1
2    SymbolRate =? 3e6 %Hz
3    EbN0 =? 12;
4    Interpolation = 16;
5    TxSampleRate = SymbolRate*Interpolation;
6    Bandwidth = 10e6;
7
8    SymbolTime = 1/SymbolRate;
9    BitsPerSymbol = 2;   % QPSK
10
11   ModPower_dBm = 13; % modulator output power in dBm
12   ModPower_W = 10^( (ModPower_dBm-30)/10 );
13   ModPower_Vrms = sqrt( 50*ModPower_W);
14   ModAmpSensitivity = ModPower_Vrms*sqrt(2);
15
16   % Eb/No = energy per bit / noise density
17   Eb_dBm = ModPower_dBm - 10*log10(SymbolRate * BitsPerSymbol);
18   No_dBm = Eb_dBm - EbN0 ;
19   NDensity = No_dBm;
```

Fig. 9.13 SystemVue Equation for BER simulation

SymbolRate should be set to the variable *SymbolRate* in Hz, *RollOff* to the variable *Rolloff*. Finally the parameter Interpolation is set to Interpolation.

AddNDensity This component adds noise density to the input. This is necessary to change the noise floor of the transmitted bits for determining the BER. In the case here considered *NDensityType* will be set to *0:Constant Noise Density*. The parameter *NDensity* will be set to the variable of same name.

Modulator In order to correctly determine the BER, not only the noise floor but also the transmitted power needs to be known. Therefore the output power of the modulator will be adjusted through the amplitude sensitivity of the modulator AmpSensitivity, which will be set to the variable *ModAmpSensitivity*.

BER_FER (Bit and Frame Error Measurement) This is the component necessary for comparing two bit streams and determining the number of errors among them. Here, the *StartStopOption* will be set to *1:Samples* and both *SampleStart* and *SampleDelayBound* will be set to 30.

At this point, the schematic should look very similar to that of Fig. 9.14. If this is the case, the simulation can almost be started. The only thing missing, in order to generate a bit curve, is to compute the BER for each desired EbNO value. For this, first you must right-click on the *Designs* folder in the *Workspace Tree* column, and then select *Add >> Evaluations >> Add Sweep* in the consecutively opening menus. A new window, entitled *Parameter Sweep Properties*, will open. In the setting, *Analysis to Sweep*, you must make sure that the schematic you have been working on is selected. In the setting, *Parameter to Sweep.* all tunable parameters will appear. Here, you must select the *EbNO* parameter, when generating the BER

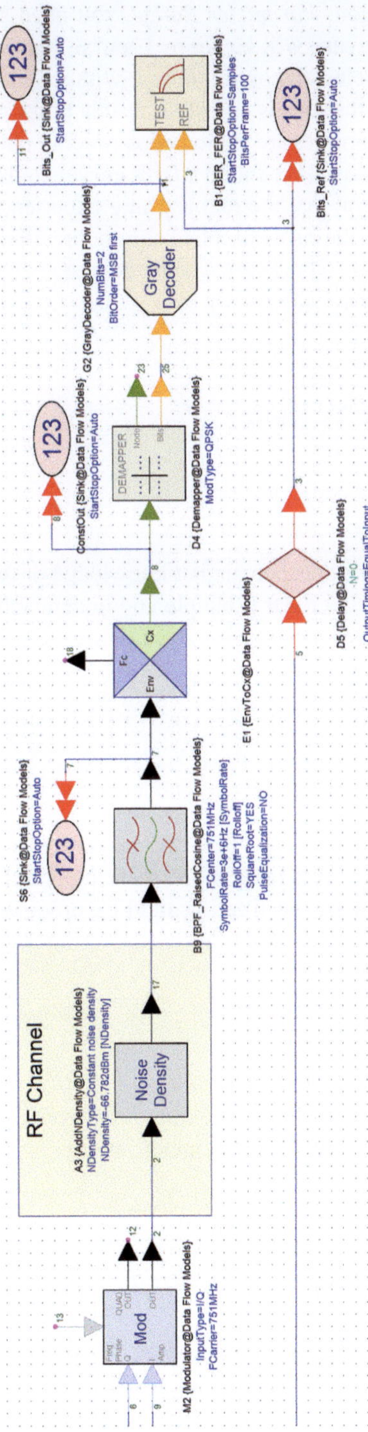

Fig. 9.14 Downconversion and demodulation chain for BER calculation

curve. Finally, in the section, *Parameter Range*, you need to adjust the *Start* and *Stop* values, which in this case will be set to 1 and 6, respectively. In the *Type of Sweep*, section you must select the option *Linear* with a Step size of 1.

The last configuration that needs adjusting at this point is the simulation time. Make sure that sufficient samples are evaluated in order to obtain good BER results. After all these steps, you must have found a BER curve similar to the typical QPSK BER curve. It is at this point that the real learning experience begins. You have at hand the possibility of simulating a complete communication system with different modulation schemes and you have learned so far how to analyze the communication system at its different stages.

9.5 Tutorial 5: FMCW Radar

9.5.1 Learning Objective

In this tutorial, the reader will deepen his/her knowledge of the FMCW (Frequency Modulated Continuous Wave) radar principle and understand the relation between the various system parameters and performance figures, like resolution and signal-to-noise ratio.

9.5.2 Prerequisites

This tutorial requires as prerequisite Chaps. 1–7 and knowledge of radar systems in general. In particular, an understanding of the working principles of FMCW radars is needed. Additional topics such as filtering and FFT processing will be considered known topics as well.

9.5.3 Design Task

In this tutorial, a frequency modulated continuous wave (FMCW) radar sensor is implemented and analyzed regarding parameter settings for an optimum signal-to-noise ratio.

9.5.4 Implementation of FMCW Radar

In Tutorial 4, we implemented a complete QPSK communication system using the Data Flow simulation engine. In this tutorial, we will use the Data Flow simulation engine once more, but it will be used to simulate a radar system, instead.

This implies the use of the *RADAR Parts* library. The radar system will consist of several stages: signal generation, up-conversion, transmission channel, down-conversion, filtering and digitization. A screenshot of the simulation model is shown in Figs. 9.15 and 9.16; a brief discussion of each stage is given below. The key radar parameters to be considered are: the signal bandwidth, the sweep duration and the carrier frequency to be set to 100 MHz, 1 ms and 24 GHz, respectively.

Signal Generation The heart of an FMCW radar is its frequency modulated signal. In general, this is done with the aid of a voltage-controlled oscillator. Within System-Vue a *Linear Frequency Modulation Waveform Generator* is utilized, instead. This is a component of the radar library that generates a linearly modulated waveform.

Up-Conversion In the last part of Tutorial 4, we learned that an RF signal in Sys-temVue is defined as an envelope signal, whereas a baseband signal is considered a complex signal, respectively. Furthermore, we saw that conversion from an envelope to a complex signal is readily available, thus avoiding the need of a mixer for signal down-conversion. This is helpful when trying to reduce the model complexity for analysis purposes. For the radar system, a simplified model will be first used and a *Complex to Envelope* component will be employed.

Transmission Channel The transmission channel is subject to different definitions. For this tutorial, the channel is considered to be comprised of both transmitting and receiving antennas as well as the propagation medium. Furthermore, these components will be modeled as a negative gain and a delay. This is, of course, the simplest interpretation of the channel, given that several effects take place when a signal propagates. For the time being, it will suffice to start analyzing this radar system. Notice that, when simulating a returned signal, the existence of a target is assumed. Thus, if there were no target there wouldn't be a received signal to be processed. Therefore, it must be taken into consideration that, if only one target exists, then there is only one received signal and, consequently, the delay and gain are correlated with the target distance and RCS (radar cross section).

Down-Conversion In the case of the radar, we will down-convert the signal with a non-ideal mixer. This way, it will possible to better experiment with imperfections in the receiving path of the radar. As the local oscillator (LO), we will use the modulated signal from the waveform generator.

Filtering Once the signal has been down-converted, adequate filtering and amplification must take place so that the radar targets can be properly recovered. Based on this, the dynamic range of the radar will be changed.

Digitization This is the final radar stage and consists in detecting the frequency of the received signal in order to resolve a certain distance and velocity. The frequency will be obtained with aid of a discrete Fourier transform which converts the time signals into frequency bins. It should be noted, however, that given the fact that only one ramp will be used, it is not possible to resolve simultaneously the velocity and distance from one single frequency measurement. Therefore, it will be assumed, unless stated otherwise, that there are no moving targets.

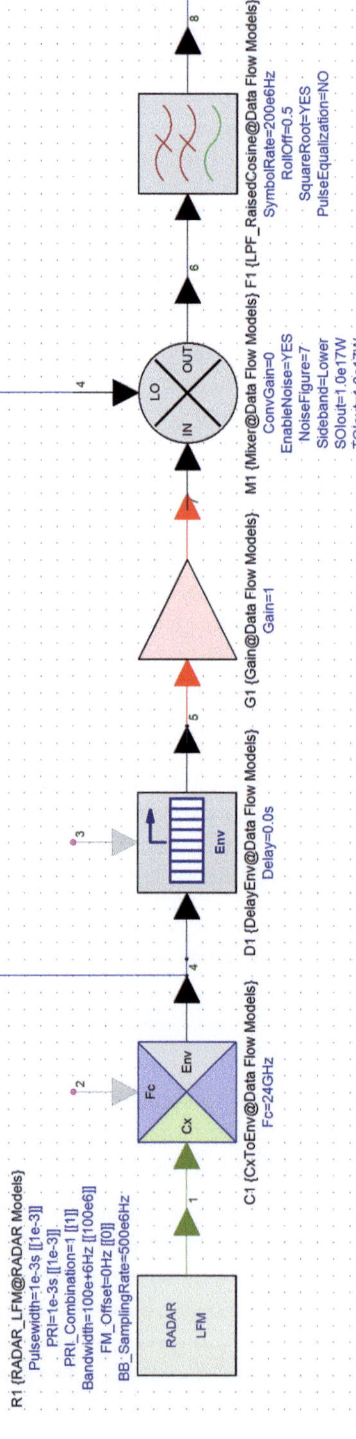

Fig. 9.15 Part 1 of FMCW Radar model in SystemVue

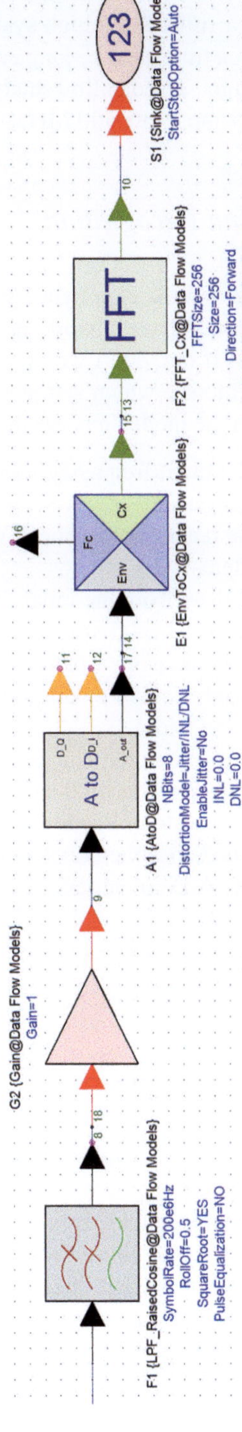

Fig. 9.16 Part 2 of FMCW Radar model in SystemVue

Simulation Settings In contrast to the previous tutorials, and given the high frequent nature of the signals used, a sufficiently high system sample rate should be chosen. Furthermore, the simulation time must be long enough to allow the signal to be reflected back from a distant target. The dataflow analysis can have, for example, a stop time of 2 ms and a system sample rate 500 MHz.

9.5.5 Target Simulation and Recognition

With the description given before and the experience you have acquired with the previous tutorials, you should already have a working model of a FMCW radar. If this is the case, congratulations! If not, don't worry, you can always take a look at the SystemVue model provided to see which settings or components you might have missed. Now, with help from the model, answer the following:

- Evaluate the received and transmitted frequency spectrum. Modify the receive filter in order to see how this impacts the received spectrum. In which way does the signal bandwidth impact the radar performance? What is the difference with respect to communications systems?
- Consider a target placed 30 m away from the radar. What time delay should be used to emulate such a target? Which beat frequency do you expect? Now, implement this delay on the model, while using a path gain of 5E-7 and observe the beat frequency at the mixer output. Compare your predicted beat frequency with the simulated one (both should coincide).
- Now, assume that the maximum measurement range of your radar should be 150 m. Given that the radar range depends on the bandwidth, undesired aliasing effects will negatively impact the system. Based on this, what cut-off frequency do you need? Which consequence does the down-sampling factor have? Which sampling rate is adequate for the A/D converter?
- Observe the radar image at the FFT output. How many peaks does an object cause in the frequency bins and which consequence does this have on the maximum measurement range?

9.5.6 Signal to Noise Ratio Analysis

A very important part of a radar simulation is determining the SNR (Signal to Noise Ratio) of the measurement. A rough estimation of the system SNR can be obtained by comparing the peak of the radar image with the mean value of all samples. One way to do this is to add a variable to the simulated dataset. Even if you haven't noticed it so far, every time an analysis is performed in SystemVue, a dataset containing all output variables is generated. In each dataset, you can see the numeric values used for generating many of the plots that have been created so far in these tutorials and, in addition, you can also generate your own variables. The definition of these variables follows the same scripting logic as with a SystemVue equation (see Tutorial 4). All you need to do is open the corresponding dataset to the radar analysis, right-click over the variables, and select the option *Add Variable*. In the window that appears, introduce the desired post-processing in the textbox labeled *Formula*. Care must be exercised, however, in case other variables using identical names are referenced. For the case of computing the SNR, a simple *max(x)/mean(x)* expression is all you need, where x should be replaced with the FFT output. Notice, however, that in order to achieve a proper result, the FFT length has to correspond to the number of samples. Now that you know how to estimate the SNR of this radar system, perform the following analysis.

- Adjust the sampling frequency, in order to achieve the full measurement range of 150 m. What happens to the sampling frequency if an IQ mixer is used instead? Compare the radar image obtained when using both mixing schemes. Are there any changes in the radar signal and in SNR? Why?
- To further improve the SNR, the ramp duration can be doubled. Change the necessary settings and observe the radar image and SNR again. Does the number of samples in the FFT change? Why does the SNR improve?

9.6 Summary

In this chapter, the student has been taken through several tutorials, which provided him/her with an opportunity to synthesize the knowledge covered in all the previous chapters of the book. The tutorials presented were conducted with the Keysight *SystemVue* electronic system-level (ESL) design software. In the tutorials, the student gathered practical experience with slightly simplified real-world design examples of radio transmission systems, both in the area of communications as well as radar sensing. The tutorials covered state-of-the-art system applications and considered the characteristics of typical radio-frequency hardware components.

Appendix

Abstract

We begin this Appendix, by presenting a rather brief summary of wireless standards. Starting with the definition of "wireless standard," we present their names and acronyms, and the tables that list their characteristics. This is then followed by a table that captures the evolution of the generations of wireless systems and their capabilities, concluding with "LTE," the Long-Term Evolution wireless system, which is the most advanced version currently deployed.

A.1 On Wireless Standards

The implementation of wireless connectivity relies upon the definition of so-called *wireless standards*. Each standard embodies the precise set of parameters that dictate the architecture and software design of wireless systems operating under the given standard, to effect compatible communication with other systems, also operating within the standard.

The typical parameters defining a standard are:

1. Multiple access
2. Frequency band
3. RF channel BW
4. Duplex method

    a. Forward/reverse channel data rate
    b. Channel coding (a system of error control for data transmission)
    c. Interleaving (a way to arrange data in a non-contiguous way to increase performance)
    d. Bit period
    e. Spectral efficiency

H. J. De Los Santos et al., *Radio Systems Engineering,*
DOI 10.1007/978-3-319-07326-2, © The Authors 2015

Table A.1 Wireless standards nomenclature [1–10]

Name of standard	Acronym
Global System for Mobile (Communications)	GSM
Digital Cellular Service	DCS
Personal Communication Services	PCS
Gaussian Minimum Shift Keying	GMSK
Complementary Code Keying	CCK
Wireless Personal-Area Network	WPAN
Gaussian Frequency Shift Keying	GFSK
Frequency-Hopping Spread Spectrum	FHSS
Direct-Sequence Spread Spectrum	DSSS
Orthogonal Frequency-Division Multiplexing	OFDM
Quadrature Amplitude Modulation	QAM
Orthogonal Frequency-Division Multiple Access	OFDMA
Multiple-Input and Multiple-Output	MIMO
Single-Carrier Frequency Division Multiple Access	SC-FDMA
Enhanced Data rates for GSM Evolution	EDGE
High Speed Packet Access	HSPA
Universal Mobile Telecommunications System	UMTS
The first cellular standard to implement the CDMA	cdmaOne, IS-85
Family of 3G mobile technology standards, which use CDMA channel access, to send voice, data, and signaling data between mobile phones and cell sites	CDMA2000

Table A.2 GSM/DCS/PCS characteristics

	GSM	DCS	PCS
Modulation	GMSK	GMSK	GMSK
Frequency band	890–960 MHz	1710–1850 MHz	1880–1930 MHz
Channel bandwidth	200 kHz	200 kHz	200 kHz
Bit rate	270 kb/s	270 kb/s	270 kb/s

Table A.3 WCDMA characteristics

	WCDMA
Modulation	QPSK
Frequency band	1920–2170 MHz
Channel bandwidth	3.84 MHz
Bit rate	3.84 Mb/s

These standards and pertinent modulation schemes are referred to by a number of acronyms, which we now give (Table A.1).

Below, a set of tables detailing the parameters for a variety of wireless standards is given [1] (Tables A.2, A.3, A.4, A.5, A.6).

Table A.4 802.11x and hyperlink 2 characteristics

	802.11FH	802.11DS	802.11a	802.11b	802.11g	HiperLAN 2
Modulation	FHSS	DSSS	OFDM; BPSK	DSSS	DSSS	OFDM
	D-BPSK	D-BPSK	QPSK	D-PSK/D-QPSK	D-BPSK/D-QPSK	FSK/ GMSK
	D-QPSK	D-QPSK	QAM	CCK	CCK OFDM BPSK/QPSK/ QAM	
Frequency band	2.4 GHz[a]	2.4 GHz[b]	5 GHz[b]	2.4 GHz[a]	2.4 GHz[a]	5 GHz[c]
Channel bandwidth	1 MHz	22 MHz	16.25 MHz	22 MHz	16.25–22 MHz	16.25 MHz
Bit rate	1.2 Mb/s	1.2 Mb/s	6–54 Mb/s	1–11 Mb/s	1–54 Mb/s	1.5–54 Mb/s

[a] 2402–2480 MHz
[b] 5150–5350 and 5725–5825 MHz
[c] 5150–5350 and 5470–5725 MHz

Table A.5 Bluetooth, ZigBee and ultra wideband characteristics

	Bluetooth	ZigBee	Ultra wideband
Modulation	FHSS; GFSK	FHSS; GFSK	OFDM
Frequency band	2402–2480 MHz	2402–2480 MHz	3100–10600 MHz
Channel width	1 MHz	1 MHz, 5 MHz	500 MHz
Bit rate	1 Mb/s	1.6–10 Mb/s	110–480 Mb/s

Table A.6 Evolution of wireless generations [2–5]

1st generation (Launched in 1980s)	2nd generation (Launched in 1991)	3rd generation (Launched in 2002)	4th generation (Launched in 2012)
Analog cellular (single band)	Digital (dual-mode, dual band)	Multi-media, multi-band software-defined radio	Multi-standard + Multi-band
Voice telecom only	Voice + data telecom (Short Text Messages)	New services markets beyond traditional telecom; higher speed data, improved voice, multi-media mobility	mobile ultra-broad-band Internet access
Macro cell only	Macro/micro/pico cell	Data networks, Internet, VPN, WINternet	
Outdoor coverage	Seamless indoor/outdoor coverage		
Distinct from PSTN	Complementary to fixed PSTN		
Business customer focus	Business + customer	Total communications subscriber; virtual personal networking	

A.2 3GPP: 3rd Generation Partnership Project [7]

The wide-band code division multiple access (WCDMA) standard has been developed for third-generation (3G) mobile systems for use in Europe and Japan. This standard was developed to offer enhanced data rates and capacity over second-generation systems. WCDMA employs frequency-division duplexing (FDD), where the transmit (Tx) signal occupies the 1920–1980-MHz band, while the receive (Rx) signal occupies the 2110–2170-MHz band. Thus, WCDMA is a full-duplex system in which the handset transmits and receives simultaneously in these separate Tx and Rx bands.

Many system performance specifications derive from this full-duplex feature. WCDMA mobile receivers require both very high sensitivity and very high linearity, with stringent out-of-band linearity requirements due to the presence of transmit leakage. Specification for key analog/RF and digital performance requirements are given in the 3GPP Technical Specification document [7].

A.3 LTE (Long Term Evolution) Receiver (4G)

The 3GPP Long Term Evolution (LTE) is the latest standard in the mobile network technology evolution that produced the GSM/EDGE and UMTS/HSPA network technologies. In fact, is has been construed as the first step towards true 4G technologies. LTE is frequently marketed as 4G, however, the first-release LTE does not fully comply with the 4G requirements [8, 9].

A truly 4G technology should be capable of download speeds of 100 Mb/s and 1 Gb/s from moving or pedestrian points, respectively [7, 8]. To avoid consumer confusion with the terms 3.5 G or 3.9 G, that began appearing, the industry decided to market LTE as "4G LTE" [8]. The main advantages of LTE are as follows:

1. High throughput
2. Low latency
3. Plug and play
4. FDD and TDD in the same platform
5. An improved end-user experience
6. A simple architecture resulting in low operating costs

LTE is also targeting the support of seamless passing to cell towers with older network technology such as GSM, cdmaOne, UMTS, and CDMA2000. A summary of the LTE capabilities is given in Table A.7.

LTE is the next major step in mobile radio communications. The advantages of LTE are high downlink date rate of 100 Mbit/s (50 Mbit/s in uplink), FDD and TDD in the same platform with simple direct-conversion architecture. LTE also should have backward compatibility to existing networks such as GSM, CDMA and WCDMA

Table A.7 Summary of LTE features

Family	Primary use	Radio technology	Downlink (Mb/s)	Uplink (Mb/s)	Notes
UMTS/4GSM	General 4G	OFDMA/MIMO/ SC-FDMA	100 (in 20 MHz bandwidth)	50 (in 20 MHz band-width)	LTE-Advanced update expected to offer peak rates up to 1 Gb/s fixed speeds and 100 Mb/s mobile users

A.4 Summary

We began this Appendix, by presenting a rather brief summary of wireless standards. Starting with the definition of "wireless standard," we presented their names and acronyms, and tables that list their characteristics. This was then followed by a table that captured the evolution of the generations of wireless systems and their capabilities, concluding with "LTE," the Long-Term Evolution wireless system, which is the most advanced version currently deployed.

References

[1] P.-I Mak, S.-P. U, and R. P. Martins, "Transceiver Architecture Selection: Review, State-of-the-Art Survey, and Case Study," *IEEE Circuits and Systems Magazine*, Second Quarter 2007, pp. 6–25.

[2] Available: [Online]: http://en.wikipedia.org/wiki/1G

[3] Available: [Online]: http://en.wikipedia.org/wiki/2G

[4] Available: [Online]: http://en.wikipedia.org/wiki/3G

[5] Available: [Online]: http://en.wikipedia.org/wiki/4G

[6] Available: [Online]: http://en.wikipedia.org/wiki/5G

[7] 3GPP TS 25.101 V5.2.0 (2002–2003) *Technical Specification* 3rd Generation Partnership Project; Technical Specification Group Radio Access Networks; UE Radio Transmission and Reception (FDD) (Release 5)

[8] Available: [Online]: http://en.wikipedia.org/wiki/LTE_(telecommunication)

[9] Available:[Online]http://www.clove.co.uk/viewtechnicalinformation.aspx?content=3B2BD491–6465-4C70-ABDB-5A12A06C3D8D

[10] Available: [Online]: http://en.wikipedia.org/wiki/CDMA2000

Index

H. J. De Los Santos et al., *Radio Systems Engineering*,
DOI 10.1007/978-3-319-07326-2, © The Authors 2015

If you have any concerns about our products,
you can contact us on
ProductSafety@springernature.com

In case Publisher is established outside the EU,
the EU authorized representative is:
Springer Nature Customer Service Center GmbH
Europaplatz 3, 69115 Heidelberg, Germany

Printed by Libri Plureos GmbH
in Hamburg, Germany